ÆDES CHRISTI
in Academia Oxoniensi
C.W.S. R.E. 1904.

Hückel Theory for Organic Chemists

THE LATE PROFESSOR COULSON
(Photograph reproduced here with permission
from Brookhaven National Laboratories)

Hückel Theory for Organic Chemists

C. A. COULSON†
BRIAN O'LEARY
University of Alabama in Birmingham

and

R. B. MALLION
Christ Church, Oxford

1978

ACADEMIC PRESS
London · New York · San Francisco
A Subsidiary of Harcourt Brace Jovanovich, Publishers.

ACADEMIC PRESS INC. (LONDON) LTD.
24/28 Oval Road,
London NW1

United States Edition published by
ACADEMIC PRESS INC.
111 Fifth Avenue
New York, New York 10003

Library of Congress Catalog Card Number: 77-93202
ISBN: 0-12-193250-8

Text set in 10 on 12 point Times by
Composition House, Salisbury

Printed in Great Britain by
Whitstable Litho Ltd., Whitstable, Kent

Foreword

Molecular-orbital theory, especially Hückel theory (1931) and its extensions, is by now widely appreciated by chemists in all areas of research and teaching. This was not always so. Actually, the method of linear combinations of atomic orbitals was used by Bloch in 1928 for wave functions in crystals. Molecular wave-functions were introduced by Mulliken (1928), Hund (1928), Herzberg (1929) and Lennard-Jones (1929) for diatomic molecules, and the extensions to polyatomic molecules followed quickly thereafter.

Charles Coulson, a student of Lennard-Jones, is largely responsible for the development and the widespread appreciation of molecular-orbital theory. He, his students, and the visitors to his research group explored many facets of this subject, and unified many new ideas over the years. In his rôles as innovator, leader in this field, and finally statesman of these parts of chemistry, chemical physics and applied mathematics, he provided a truly creative atmosphere for his research group. For many years there was no comparable place for the study of molecular theory of complex molecules, and for that reason the very wide interest that now exists in molecular-orbital theory can be attributed to Coulson's imagination and teaching.

I am sure that he, a modest and generous person, would register some protest to these observations. Indeed, he always emphasised a balanced view even when allegiances to valence-bond theory *versus* molecular-orbital theory were almost divisive. His own words were,

> "It has frequently been the custom for supporters of one or the other of these theories to claim a measure of chemical insight and quantitative reliability for the method of their choice. This is a pity because neither method is complete or fully satisfactory. Fortunately in most of their conclusions the two theories agree, although they reach their conclusions in quite distinct ways. This means that we must not arbitrarily reject one or the other. . . . There is little doubt but that thc molecular-orbital theory is conceptually the simplest. Historically it was developed after the other theory was already established, and for that reason has been a little slower in gaining acceptance. Its present status, however, in dealing with excited states, is fully equal to that of the valence-bond theory."

The book, *Valence* (Clarendon Press, Oxford, 1952) from which this quotation was taken, is a masterpiece of chemical and physical intuition, written with a simplicity and clarity fully characteristic of his lectures, which I was privileged to hear during my sabbatical year at the Mathematical Institute in Oxford in 1954–5.

His own original work in this area began with the first quantum-mechanical study of a three-centre bond (H_3^+) in 1935, and continued almost uninterrupted until early 1974. For most of this period he was a major focus of developments and applications of theoretical chemistry. Among his many original contributions the early conception, development and application of bond orders remain as my choice of his best and most influential research. The close correspondence between interatomic distances predicted in aromatic hydrocarbons compared with the X-ray results, mostly from J. M. Robertson's laboratory, transformed molecular-orbital theory from a field of interest to specialists into a subject of widespread interest to all chemists. The related ideas on charges, free valence and chemical reactivity produced a relevance of theory to experiment not previously realised. Coulson's intuition is well exhibited here in his emphasis on molecular properties which are largely independent of both the total molecular energy and the parameters of Hückel theory itself. Also closely related are the ingenious papers with H. C. Longuet-Higgins on electronic structures of aromatic hydrocarbons, and with W. Moffitt on strained bonds and hybridisation in hydrocarbons. Three lines of Coulson's work influenced me greatly. The first was his study of multicentred integrals with M. Barnett which led us to the first accurate calculation of the barrier to internal rotation in ethane (with R. M. Pitzer). In the second, both our interest and his beyond π-orbital theory encouraged us to programme the extended Hückel method (with R. Hoffmann and L. L. Lohr, Jr). The third, our comparison of the Edmiston–Ruedenberg and Boys procedures for localisation of orbitals, was dedicated to his memory in 1974.

This book, an account of Coulson's lectures from recordings and notes by Brian O'Leary and R. B. Mallion, is far more than a historical document. One can perceive the way in which Coulson thought, his mixture of optimism and pessimism, his sense of rigour and of approximation, and his guesses and analyses. Moreover, one can sense his dedication to the specific problems, and most of all to his communication of an underlying physical and chemical intuition. While the quantitative aspects of molecular-orbital theory have now gone well beyond the Hückel method, modern computer techniques have not replaced the simple, intuitive ideas of molecular-orbital theory, best exhibited by Hückel theory as developed in this small volume.

William N. Lipscomb,
Harvard University,
Cambridge, Massachusetts, U.S.A.

July, 1978

Preface

Chemists are rarely sentimental, particularly about other chemists. As is the case within any branch of science, the professional chemist finds himself in a world of rivalry (good-natured or otherwise) all too often polarised around personalities. To this world, C. A. Coulson brought a quality of vision, warmth and kindness that only those who knew him could appreciate. We feel that to have been members of his group at the Oxford University Mathematical Institute, and then to have moved with him to his Theoretical Chemistry Department, represents the rarest of privileges—speak to any "Coulson-Chemist" and you will evoke the same response.

"But what was he really like?", one can almost imagine future generations asking. We hope that this book will answer, at least in part, that question. By, as-it-were, eavesdropping on his lectures, we hope that something of the flavour of an era, now sadly past, will return.

How were we able to do this? Early in 1971, one of us (B.O'L.), with C.A.C.'s permission, tape-recorded the Professor's lectures to the third-year undergraduate-chemists, mainly because we felt that we should like to have a record of Hückel Theory taught by the man who had virtually created it in its present form.

Why edit and publish this material? Shortly after Professor Coulson's death (which occurred on January 7th, 1974), a fund was set up in his name with the object of bringing to Oxford students who otherwise could not afford to come, to study or research in Theoretical Chemistry. Both of us have benefited so enormously from our Oxford years and from our work with "the Prof." that we decided to prepare his lectures for publication, with the proviso that all royalties from the sale of the work go to the Coulson Memorial Fund.

The late Professor Coulson gave his course of eight lectures entitled "Hückel Theory for Organic Chemists" to the third-year undergraduates of Oxford University in 1967, 1968, 1969, 1971 and 1973. The Professor also started giving these lectures during his final illness in Michaelmas

Term 1973, but these had to be discontinued when his health broke down in the November of that year, and this last course was in fact completed by one of the present authors (R.B.M.). This lecture series, though quite extensive in its content and by no means superficial in its treatment, was tailor-made for its undergraduate recipients, the only mathematics assumed (mainly a rudimentary knowledge of the theory of matrices and linear equations) being taught in the first-year "Mathematics for Chemists" course at Oxford.

It should of course be emphasised that C.A.C.'s fundamental and original contributions to Hückel Theory (which could almost, perhaps, with some justification, even be called "Coulson Theory") are well documented in the learned journals from 1939 onwards. Perhaps because of this, the Professor never did set down completely, in written form, the way in which he expounded this subject, in a detailed and rather leisurely fashion, to undergraduates. It was in the Spring of 1974, after the Professor's death, that Mrs Eileen Coulson consented to the idea that, if all the 1971 tapes could be located, and if the recording quality were adequate, they should be transcribed and edited for publication. This would achieve two purposes: it would preserve for posterity the way in which C.A.C. put this subject over to an undergraduate audience and, we hope, it will provide some welcome income for the Oxford University Coulson Memorial Fund.

By the summer of 1974, the tapes of all eight lectures had been found (that of the first lecture being particularly elusive) and, between November 1974 and January 1975, each one was faithfully and entirely transcribed by B.O'L. who made use of an extensive tape-amplification system available at the University of Alabama. It should also be pointed out that complete notes of all diagrams and equations drawn on the blackboard were taken at the time that the lectures were delivered in 1971, and Mrs Coulson generously provided a copy of the Professor's own notes for the course, from the Coulson Archives. During the Long Vacation of 1975, the Oxford-based author (R.B.M.) went to Birmingham, Alabama, and, on the basis of this exceptionally complete documentation of the 1971 lectures, produced the edited version of them presented in this book. (R.B.M. is very grateful to Dr J. Macey of the Computing Department of the University of Alabama in Birmingham, the Royal Society of London, the Oxford University Lockey Bequest, and the Governing Body of Christ Church, Oxford, for their financial assistance which made this visit possible).

Having said the above, however, we must put on record the fact that the text is by no means claimed to be a *verbatim* account of the lectures; although an impressively large proportion of the Professor's spontaneously spoken sentences stand up remarkably well, almost unchanged, in print—being, as they are, accurately grammatical and pleasingly constructed—there has inevitably been the necessity for extensive revision of the presentation.

Consequently, while most of the text is, we hope, still very much in the spirit of Coulson's original delivery, we have, in many cases, felt it appropriate even to reorganise the actual *order* of the presentation of certain material, as well as to expand it in some places, and contract it in others. Further additions involve the use of Graph Theory, the mathematical discipline which underlies this simplest-of-all molecular-orbital method and which, in our opinion, serves to emphasise the basic elegance of Hückel Molecular-Orbital Theory. Some Graph Theory is therefore used in the text, but the major part of it is banished to Appendices (A, C, and D) which are entirely due to R.B.M. and B.O'L.

In order to place on permanent record the precise contribution to this book (and it is, of course, the major one) made by the late Professor Coulson, copies of the actual recordings will be deposited with the Coulson papers in the Contemporary Scientific Archives Collection of the Bodleian Library, University of Oxford, and it is hoped that the transcripts will also be available, either in the above Collection or *via* Mrs Coulson's personal archives.

We should like to devote our closing remarks to an expression of our sincere thanks to Mrs Eileen Coulson for all the many kindnesses she has shown us both over the years, and to Miss Rosemary Schwerdt, "the Prof.'s" secretary for 19 years, who has generously given of her time and talents in helping us to prepare the manuscript of this book.

BRIAN O'LEARY
Birmingham, Alabama, U.S.A.

June, 1978

R. B. MALLION,
Canterbury, England, U.K.

Copyright Acknowledgments and Disclaimer

By the very nature of this book, and the unusual circumstances of its production, there have been some difficulties in establishing the original sources of diagrams and figures which, having been displayed on the blackboard or projected as slides when Professor Coulson gave the lectures, were copied down and have been redrawn for the present volume. In some cases, we have been able to trace the owners and permission to reproduce the figures in question has been sought, obtained, and is gratefully acknowledged below. Most of the other diagrams, we are therefore forced to conclude, must be the Professor's own creation. There may, however, be some figures which we have reproduced, in all good faith, without acknowledgment to, or permission from, the original source, simply because this has not been known to us. Apologies are therefore offered in advance to the authors, publishers, or other copyright-owners of any such material and we can only ask for their kind indulgence and understanding in this matter.

Concerning those figures the owners of which we *could* trace, we thank the following:

For FIG. 4-5 Oxford University Press

FIG. 4-6 Dr A. J. Taylor (*née* Buzeman) and the Physical Society of London

FIG. 4-11 Dr E. C. Kooyman and the Chemical Society of London

FIG. B-2 Professors H. Eyring, D. Henderson and W. Jost.

We should also like to state, in conclusion, that for Appendix B, and some of the material in Chapter IX, we have made more than casual use of C. A. Coulson's chapter on "π-Bonds" in *Physical Chemistry: An Advanced Treatise*, Vol. V *Valency* (H. Eyring, D. Henderson and W. Jost, Eds), Academic Press, New York, 1970, in which C.A.C. had discussed similar subject-matter. We are very grateful to the Editors for permission to do this.

Contents

One

Foundations of Hückel Theory

1.1 Historical Introduction

The discussion which we shall be concerned with in this book began in 1931, quite soon after Wave Mechanics had been introduced into Chemistry, with the work of Erich Hückel; for that reason, the subject is often referred to as Hückel Theory. It was, in its earlier stages, a very *ad hoc* affair—perhaps one could say that it still is. If, however, we consider some figures quoted by Streitwieser it is evident just how the subject has mushroomed in the intervening 40 years. In the 1930's, according to Streitwieser, approximately 20 papers were published in this field[N1]*; in the 1940's there were *ca.* 70 papers; in the 1950's there were *ca.* 600 papers; and an extrapolated figure for the 1960's reveals that something of the order of 5000 papers concerned with Hückel theory were published during that latter decade. Furthermore, there seems to be no sign of this trend's stopping, even though, as we shall see before the end of this discussion, the treatment in most papers is now becoming more sophisticated.

Here, therefore, is a subject which has been exerting a fairly considerable influence on the way in which organic chemists express their conclusions and describe their work. It rests upon one fundamental distinction—that between all σ- and π-electrons; it is to this distinction that we now turn.

1.2 The idea of σ-π Separation

A general, unsaturated molecule is a planar, or near-planar, system; for the moment we shall consider it to be exactly planar. In this very special situation, molecular orbitals describing the molecule as a whole will inevitably divide themselves into two fundamentally different types. These types are distinguished by symmetry—those which are symmetric and those which are

* Notes, denoted by the superscript N, are listed on p. 167.

anti-symmetric with respect to reflection in the molecular plane. Consider, for example, a spherically symmetric type of atomic orbital such as a 1*s*-orbital on carbon or hydrogen; this is quite obviously unchanged on reflection in the plane. A similar situation is encountered if we consider a *p*-orbital *provided that* its axis of symmetry lies completely in the plane, the parts of it above and below the plane again being unchanged on reflection. However, consider a *p*-orbital which has its axis of symmetry perpendicular to the molecular plane, like that shown, for example, in Fig. 1-1. Here, if the

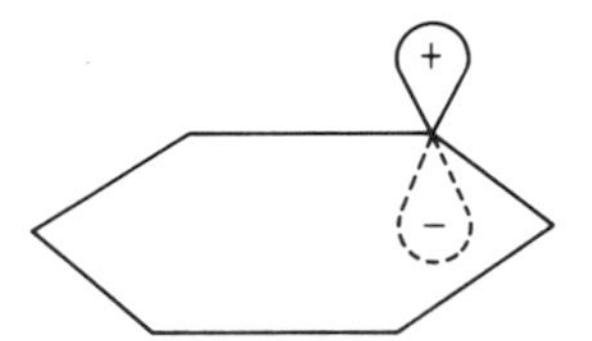

FIG. 1-1 p_π-atomic orbital on a carbon atom in benzene

analytical expression for this orbital is positive above the plane, it must be negative below the plane; hence this wave function is of the type that is anti-symmetric with respect to reflection in the molecular plane. Furthermore, as will be seen later, mixing between the symmetric and anti-symmetric types of function will be nil. Consequently, if we were to form a molecular orbital by combining together a number of individual atomic orbitals—of whatever kind they may be—then those of the one symmetry-type will never be combined with those of the other. Immediately, therefore, we are led to the distinction between functions (designated σ) which are symmetric when reflected in the molecular plane, and those (called π) which are anti-symmetric with respect to the molecular plane. It is these so-called π-functions which are the interesting ones from the point of view of the present discussion, though we shall have something to say later on about the σ-orbitals, and the interaction between σ and π.

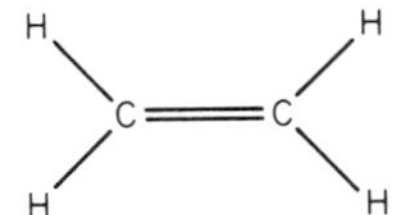

FIG. 1-2 Structural formula for ethylene.

Let us illustrate these ideas by discussing the ethylene molecule (Fig. 1-2) in some detail. The two carbon-atoms are considered to be in a state of sp_2-trigonal hybridisation. By using an arrow to represent the orientation in which these sp_2-hybrids are directed, the carbon hybrids may be represented as in Fig. 1-3. These hybrids are all of the σ-type and so it is natural to suppose that bonds can be made from them. First of all, we can consider the formation

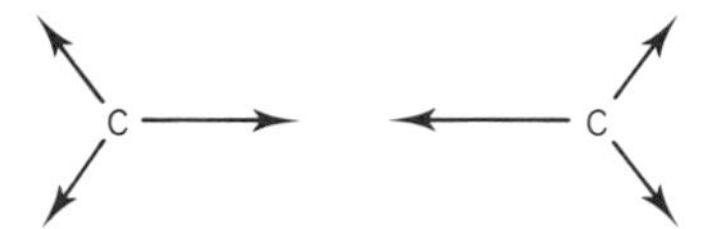

FIG. 1-3 Orientation of carbon sp_2-hybrids in ethylene.

of a carbon-carbon bond by pairing two of these hybrids together (either in terms of a molecular-orbital (MO) or valence-bond (VB) wave-function—at this stage it does not matter which we consider it to be) as in Fig. 1-4.

FIG. 1-4 Pairing of two carbon sp_2-hybrids to form a single carbon-carbon bond in ethylene.

Carbon-hydrogen bonds can then be formed by attaching hydrogens to the remaining sp_2-carbon-hybrids, as is shown schematically in Fig. 1-5. Each one of these four hybrids overlaps very effectively with its neighbouring hydrogen, but with very little else. Consequently, we could regard the properties of the C—H bonds as being very largely determined by a given carbon-sp_2-hybrid and the 1s-orbital of the adjacent hydrogen. Equally well, the σ-part of the double bond is controlled by the two carbon-hybrids involved and by very little else and so we can speak of these as being localised; however, when we come to consider the remaining orbitals (in larger systems, particularly) this is not going to be true. We have now accounted for the s-orbital of the carbon valence shell and the p-orbitals which lie in the plane; there now remains the third p-orbital (often referred to as a p_π-orbital

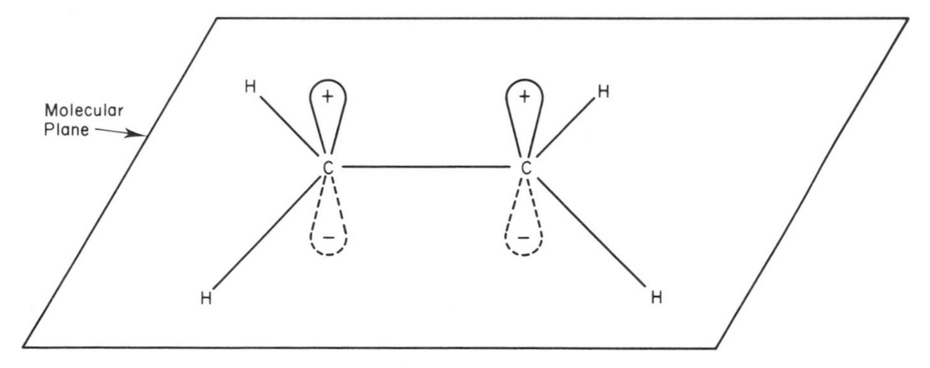

FIG. 1-5 Carbon-carbon and carbon-hydrogen σ-bonds, plus (non-interacting) carbon p_π-atomic orbitals, in ethylene.

for the symmetry reasons outlined above); this is the *p*-orbital illustrated schematically in Fig. 1-1. The two p_π-orbitals for ethylene are shown in Fig. 1-5; these two orbitals can now be paired together. In Fig. 1-5 they have been drawn as though they were very much separated from each other but that is only for convenience in the diagram. If we now imagine these two orbitals overlapping there will be an overlap region above and below the plane of the molecule, and there will then be what we could regard as a bond formed, with its maximum charge-density *above* and *below* the plane—*not in* the plane. So we see that this new sort of bond is quite different in that respect from the σ-bond. However, it is the *symmetry* which is the important factor here; the symmetry is such that the wave function changes sign on reflection in the plane of the molecule whereas in the case of the σ-bonds referred to above the wave function does not change upon such a reflection.

Now suppose that we had not two but three or four of these atoms in a row (Fig. 1-6). This is the situation we might imagine, for example, with butadiene. If we consider the p_π-orbital on, say, the second of these atoms, labelled 2 in Fig. 1-6, then this could overlap, for example, with the orbital on carbon-atom 1—or, it may equally well overlap with the p_π-orbital on carbon-atom 3; similarly, 3 can overlap equally well with 4 as with 2. It is therefore no longer rational to suppose that we have here a localised bonding in the way in which it was reasonable to postulate with the σ-electrons; in fact, in this model, the molecular orbital formed from the overlap of the p_π-atomic-orbitals would be expected to extend over all four carbon atoms of the molecule; it may not extend equally over all of them, but that is something which, for the moment, we have to discount. The main feature to be

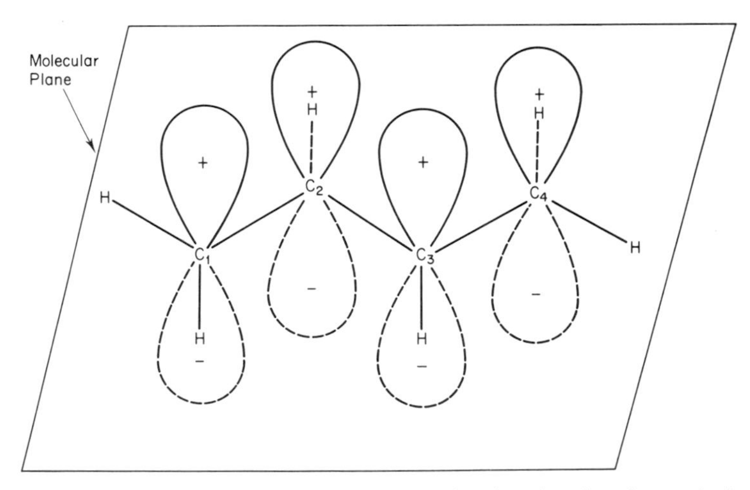

FIG. 1-6 Carbon-carbon and carbon-hydrogen σ-bonds, plus (non-interacting) carbon p_π-atomic orbitals, in butadiene.

recognised here is that provided the p_π-atomic-orbitals in question are parallel with one another, or at least are closely parallel, we shall expect to have what we might call *delocalised bonding* in the sense in which such delocalised bonding is *not* displayed by the σ-electrons.

That is a very simple picture but it really has the bones of the matter in it. It explains, for example, how it is that if we were to make some alteration at an end carbon-atom—perhaps putting a chlorine atom or fluorine or some other substituent instead of a hydrogen—then an influence of this change is felt in the most distant parts of the molecule; such an effect would not be anything like so pronounced in the case of the σ-bonds and we see that it is intimately bound up with the π-electrons. It is well known that these effects do occur—substitution reactions in substituted benzenes and so on—and it appears that this is the basis in terms of which these effects ought to be discussed. But it all depends on symmetry.

1.3 Other Specific Properties of Conjugated π-Electron-Systems

We have other reasons for being interested in these π-electrons. Above, we have referred to the orbitals of one atom overlapping more or less equally with the orbitals of adjacent atoms leading to what we have called delocalised bonding. However, we can also make reference to bond lengths which are characteristic of conjugated systems of the type being considered involving this delocalised bonding, and which distinguish such bonds from the localised single- and double-bond types. For example, in benzene, the C—C bond-length is close to 1·40 Å, whereas a C—C single-bond in ethane is of length 1·54 Å and an olefinic double-bond such as occurs, for example, in ethylene, is 1·34 Å. The benzene bond-length is thus something in between a single and a double bond. What is more, all such bonds in benzene are equal; in polycyclic systems like naphthalene, for example, (Fig. 1-7) the

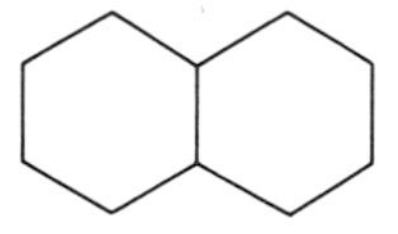

FIG. 1-7 Carbon-atom σ-bond connectivity of naphthalene.

lengths of chemically non-equivalent bonds do in general differ from each other—though they are still clustered around the benzene value of 1·40 Å. Indeed, some of the earlier books (*e.g.* Sidgwick's[R1*]) refer to this as the

* References, denoted by the superscript R, are listed on p. 172.

"aromatic bond-length". This kind of language is sometimes useful but we shall see later (Chapter Four) that C—C bond-lengths are not by any means the same for all so-called aromatic systems.

Another feature of these conjugated molecules which calls for explanation is their characteristic ultra-violet (UV) properties. On the whole, single bonds absorb energy at rather a short wave-length (*i.e.* high energy), but with the type of molecule we are discussing absorption is often obtained at much lower energies, and that leads to a way of characterising this kind of molecule. One finds, for example, that there are interesting sequences of absorptions if we start with benzene and then condense a further ring to obtain naphthalene and then continue the annellation through anthracene, tetracene, etc. Another important aspect of these molecules (which has even been suggested as a basis for their characterisation) is their notable magnetic properties which manifest themselves in large magnetic-susceptibility anisotropies and very low-field (or, occasionally, very high-field) proton-chemical-shifts in a proton-NMR experiment. Then there is the fact that in systems of this sort a substitution reaction is nearly always preferentially favoured over an addition reaction; this, of course, opens up the large field of aromatic substitution.

1.4 Valence-Bond (VB) *vs*. Molecular-Orbital (MO) Formalisms

These, then, are some of the properties which are experimentally observable and for which an explanation is needed. The first clue to such an explanation has already been given in our agreeing to discuss particularly the π-electrons of such systems, but we must now decide on the way in which we are henceforth going to deal with them. We are going to use the Molecular-Orbital (MO) method for most of the discussion in this book. The MO-method will be seen to have the great advantage here that it builds in the required delocalisation right from the start. Now, of course, use of the MO-method is not mandatory and, indeed, it was not nearly so fashionable in early work as it is now. In the case of benzene, for example, people preferred to go back to Kekulé's original idea in which one imagined what in present-day language would be called "resonance" between the two "Kekulé" structures

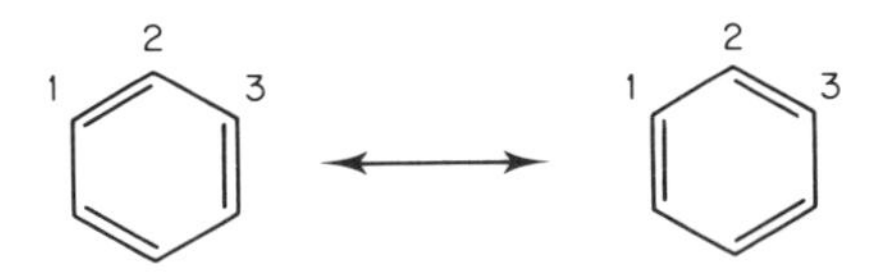

FIG. 1-8 The two, equivalent ("unexcited") Kekulé-structures for benzene.

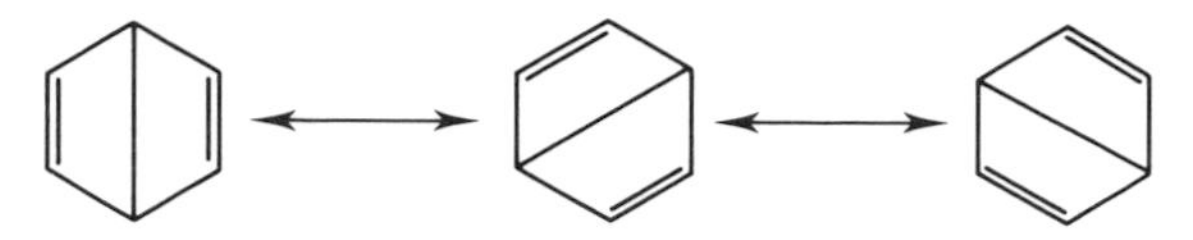

FIG. 1-9 The three, equivalent, "singly-excited-Kekulé" (or "Dewar") forms of benzene.

(Fig. 1-8)—and, of course, in addition, the three "Dewar" structures (Fig. 1-9). In a sense one could interpret this as implying "delocalisation" because we might say that, in the first Kekulé-structure of Fig. 1-8, the π-orbital on carbon-atom 1 is paired with the one centred on carbon-atom 2 and so in a sense they are sharing that region of space between them and have it, so-to-speak, "to themselves". A similar point could be made for other formally "localised double-bonds" in the various Kekulé-structures so that, in a rather anthropomorphic way of talking (which must be treated with considerable caution!), one can imagine two electrons "changing places" so that, for example, they are "sometimes" in the region between atoms 1 and 2, and "sometimes" in the region between atoms 2 and 3. One may, if one wishes, think of this as "delocalisation", though the reader may be warned that this is a particularly incautious and sometimes dangerous way of thinking! Whilst this method scored heavily in the early days of Quantum Chemistry, mainly because of the influence of Pauling and Wilson, it has now almost completely gone out of fashion[N2]. This is why, in the subsequent discussion, we are going to use the MO-method.

Two

The Hückel Molecular-Orbital Method

2.1 Linear-Combination-of-Atomic-Orbitals (LCAO) Approach used in MO-Theory

In this book we shall adopt the Linear-Combination-of-Atomic-Orbitals (LCAO) form of molecular-orbital theory. That is to say, we suppose that each of the π-electrons has a description which allows it to move over the whole framework of the conjugated system. This description allows a molecular orbital, Ψ, to be expressed

$$\Psi = \sum_{r=1}^{n} c_r \phi_r \tag{2-1}$$

in terms of a weighted (c_r) linear-combination of atomic orbitals, ϕ_r; in equation (2-1) the sum is over the running-index, r, which successively designates in turn the n different atoms which comprise the conjugated system. In the case of benzene, for example, there would be six atomic orbitals to consider in the simplest LCAO-form, and so n in equation (2-1) would be 6; later, we shall see how to extend the "basis" (as the set, $\{\phi_r\}$, of atomic orbitals out of which Ψ is constructed, is called). This simplest basis (one atomic orbital per atom in the conjugated system) is called a "minimal basis" and it is the one which we shall use in most of our discussions in this book.

It is evident from the above that an LCAO-MO is considered to have some appropriately weighted characteristic of, or contribution from, each of its constituent atomic orbitals. Given (or, rather, assuming) the set, $\{\phi_r\}$, our next task is to find appropriate values for the weighting-coefficients, $\{c_r\}$; this was the question to which Hückel addressed himself in 1931. Of course, there will be a different set of n coefficients, $\{c_{Ir}\}$, for each different MO[N3], Ψ_I. Strictly, therefore, "Ψ" in equation (2-1) should be "Ψ_I", since there is more than one combination of the given atomic orbitals which will form a suitable molecular orbital, and "c_r" should, consequently, be "c_{Ir}"; for the moment we shall drop the subscript, "I"; we shall, however, reinstate it later.

This LCAO-approach is, of course, an approximation and is not in general going to yield the best-possible molecular orbitals. When the latter are attained one is said to have reached the *Hartree–Fock Limit*; coming close to this limit leads one into some very heavy (though interesting) calculation and we shall not attempt to do this in the present discussion. We shall see, however, that it is still possible to obtain much useful information without going to the Hartree–Fock Limit.

2.2 The Variation Method

Even if we accept that we cannot go to the Hartree–Fock Limit the question of establishing an appropriate and effective criterion for determining the combinatorial coefficients, $\{c_r\}$, of equation (2-1) still remains with us. For this purpose, the criterion which has been used is called the *Variation Method*. This is a fundamental principle of Quantum Theory which we merely adopt here without proof. According to this principle, the $\{c_r\}$ are chosen in order that a certain quantity (called the "Rayleigh Ratio" because it is derived chiefly from the work of Lord Rayleigh at the beginning of the present century) shall be stationary with respect to variation of the $\{c_r\}$.

In order to obtain the Rayleigh Ratio it is necessary to introduce the concept of an "effective" wave-equation for each electron (specifically, for each π-electron since, for the moment, these are all we are discussing). If there really were only *one* electron which we had to deal with there would be no problem since the Schrödinger Equation for a single electron is well known and can immediately be written down. There is a very convenient shorthand-symbolism for this and, for one electron—which of course is not a very realistic situation in organic chemistry—the wave equation can be written in the form

$$\mathscr{H}\Psi = E\Psi \tag{2-2}$$

in which Ψ is the wave function of the single electron in question, E is its energy and $\mathscr{H}$ is a differential operator[N4] referred to as the Hamiltonian; this is the sum of the kinetic- and potential-energy operators for the system and for a single electron is given by:

$$\mathscr{H} = \underbrace{\frac{-h^2}{8\pi m}\nabla^2}_{\text{kinetic-energy operator}} + \underbrace{\mathscr{V}}_{\text{potential-energy operator}} \tag{2-3}$$

in which $\mathscr{V}$, the potential-energy operator, is a simple multiplicative operator and ∇^2 is the standard (second) differential ("Laplacian")-operator, more explicitly written

$$\nabla^2 = \frac{\partial^2}{\partial x^2} + \frac{\partial^2}{\partial y^2} + \frac{\partial^2}{\partial z^2} \tag{2-4}$$

There is, therefore, no problem in writing the wave equation for a single electron; but if there are several electrons then what is really required in the Schrödinger wave-equation is a kind of global Hamiltonian in which all the electrons appear. We can at least write down its form. Firstly we shall require an operator corresponding to ∇^2 to act on the set of coordinates of each of the several electrons—*i.e.* ∇_i^2 where i runs through $1, 2, \ldots, n$ giving ∇_1^2, ∇_2^2 etc., according to the number of π-electrons we have in the system under investigation. Then there are a large number of terms arising in the potential energy. If, for example, we were dealing with a 10-π-electron system such as naphthalene the resulting expression would not be very elegant or manageable; but it would nevertheless be the correct π-electron-Hamiltonian, even though we would not be able to make much progress with it in a simple way. What we want to do, therefore, is to replace this correct, but complicated, Hamiltonian involving several electrons by an *effective* Hamiltonian for each electron separately. If we were not able to do this (*i.e.* replace the full Hamiltonian for many π-electrons by a separate Hamiltonian for each) then Hückel theory would probably never have got off the ground at all.

One naturally now asks what that Hamiltonian would be if it were written down (although, in fact, it is usually *not* written down!). We shall designate this Hamiltonian as $\mathscr{H}_{\text{effective}}$ in order to make it quite clear that it refers to *one* particle, (one of the π-electrons); it can then be written in the form

$$\mathscr{H}_{\text{effective}} = \frac{-h^2}{8\pi m}\nabla^2 + \mathscr{V}_{\text{effective}} \tag{2-5}$$

This illustrates the major conceptual difficulty that arises in this approach, *i.e.*, the term $\mathscr{V}_{\text{effective}}$. If, for example, a two-electron system were being dealt with then these two electrons (call them 1 and 2) exert a repulsion on one another and so in the expression for the total potential energy there must be a term e^2/r_{12}, where r_{12} is the momentary separation between the two electrons. Such a quantity cannot, of course, be brought into any expression which claims to be a one-electron (full) Hamiltonian for electron 1, for this makes no reference to the coordinates of electron 2—and therein lies the difficulty. In adopting this *effective*, one-electron Hamiltonian, therefore, it is evident that we have in some way implicitly assumed some kind of disguised, hidden interaction between the several electrons in the system—*i.e.* we have *averaged* these interactions. In doing this, we are going to lose the sort of intimate correlation between electrons which, for example, enables us to say crudely that if electron 1 is momentarily in one part of the molecule, the other electron, 2, is more likely to be in a part of the molecule distant from 1. One therefore, in a sense, "pays" for the use of an effective Hamiltonian in this way but at least when this approximation *has* been adopted the resulting Hamiltonian *can* be conveniently handled, as will be seen. In spite of this last statement, however, to write down in practice precisely what this Hamiltonian would be is a very difficult job indeed, and it will not be attempted here. In fact, the great merit of the Hückel theory is that, in most cases, we do not *need* to know explicitly what $\mathscr{H}_{\text{effective}}$ is! What we shall discover presently is that the results of calculations in Hückel theory are given in terms of one or two empirical parameters (conventionally called α and β) and, within the degree of validity of this discussion, it is not necessary to know exactly what the numerical values of these parameters are. Trends in properties along a set of related molecules, as well as ratios of quantities expressed relative to corresponding calculated quantities for (say) benzene, may be obtained which are independent both of the precise numerical values of α and β and of the algebraical form of the effective potential, $\mathscr{V}_{\text{effective}}$. We shall from now on, therefore, concentrate on $\mathscr{H}_{\text{effective}}$ without knowing precisely what it is! Now this is in some ways a rather unsatisfactory aspect of the Hückel theory but once it is accepted (and the reader is asked to accept it, or there is no point in his reading further!) the subject develops very elegantly, as will be seen. This procedure has the consequence that instead of solving an equation like (2-2) in which $\mathscr{H}$ is a many-electron Hamiltonian with an impracticably large number of terms, we solve only equations like (2-6), which contain just *one* set of coordinates dealing with each electron separately. That is to say, our job is to solve:

$$\mathscr{H}_{\text{effective}}\Psi = E\Psi \qquad (2\text{-}6)$$

The reader might of course object that it is going to be very difficult to solve an equation of this sort when we do not even know what $\mathscr{H}_{\text{effective}}$ is; that would be fair comment. This difficulty can, however, be at least partially avoided by the following process: we can go quite a long way by substituting equation (2-1) for Ψ and then keeping certain quantities vague as long as possible during the calculation and manipulation, and only at the end finally seeking some experiment which will give numerical values to the parameters which this procedure has forced us to introduce. We now examine how this philosophy works in practice.

Now that the idea of an effective Hamiltonian has been fully expounded, we may return to our discussion of the Variation Method and the Rayleigh Ratio. For simplicity, in the remainder of this discussion the subscript "effective" will be omitted from the symbol, "$\mathscr{H}_{\text{effective}}$"; hence in all that follows, $\mathscr{H}$ (which, in equations (2-2) and (2-3), represented a rigorously correct, π-electron Hamiltonian) will be understood to stand for an *effective, one-electron* Hamiltonian.

The Rayleigh Ratio, denoted by y, is defined as:

$$y = \frac{\int \Psi^* \mathscr{H} \Psi \, d\tau}{\int \Psi^* \Psi \, d\tau} \tag{2-7}$$

Since Ψ is given, in the LCAO-MO approximation, by equation (2-1), it follows that:

$$y = f(c_1, c_2, \ldots, c_n) \tag{2-8}$$

It will be recalled that the Variation Principle asks us to choose those values of $\{c_r\}$ which make the Rayleigh Ratio, y, stationary. We thus seek a situation in which a change in c_r ($r = 1, 2, \ldots, n$, in turn) produces no change in y. In other words, we shall require n conditions of the sort

$$\frac{\partial y}{\partial c_r} = 0, \qquad r = 1, 2, \ldots, n \tag{2-9}$$

to hold. This process will yield sets of coefficients, $c_1, c_2, \ldots, c_n$ which, in the present model, will represent the best that we can do.[N5] Our hope is that if we have a similar level of approximation at all stages of the calculation much the same kind of error might arise at all these stages in calculations on *different* molecules so that comparisons between one system and another will still stand out.

Proceeding in this rather optimistic frame of mind we firstly have to find out what the Rayleigh Ratio is—*i.e.* we shall have separately to discuss the numerator and the denominator of the expression in equation (2-7).

The $\{\phi_r\}$ in the expression for Ψ

$$\Psi = c_1\phi_1 + c_2\phi_2 + \cdots + c_r\phi_r + \cdots + c_n\phi_n \tag{2-10}$$

(where there are n atomic orbitals in the molecular framework) are atomic orbitals which, it is assumed, are known. If benzene were being considered then they would be the $2p_\pi$-orbitals of each carbon atom; if a calculation were being performed on pyridine then five of $\{\phi_r\}$ would be carbon-$2p_\pi$ and the sixth would be a nitrogen-$2p_\pi$. Firstly, we assume that these basis orbitals are normalised—*i.e.*

$$\int \phi_r^* \phi_r \, d\tau = 1 \qquad (r = 1, 2, \ldots, n) \tag{2-11}$$

(Since [unless a magnetic field is present] $\{\phi_r\}$ and Ψ can always be chosen to be real, complex-conjugates will, from now on, be dropped from expressions such as (2-11), the latter equation becoming

$$\int \phi_r^2 \, d\tau = 1. \qquad \text{(2-11a))}$$

We also make another assumption, at the moment simply for convenience. It is an assumption which Hückel made and one that has been made by a very large number of people subsequently. We invoke it partly on historical grounds and partly because it is simple; afterwards (§2.3) we shall go back and see what difference it has made. The assumption is that atomic orbitals on different centres are orthogonal—*i.e.*

$$\int \phi_r \phi_s \, d\tau = 0, \qquad r \neq s \qquad \begin{matrix} (r = 1, 2, \ldots, n) \\ (s = 1, 2, \ldots, n) \end{matrix} \qquad \text{(2-12)}$$

Now this is certainly not true if we were to take, for example, two adjacent carbon atoms in ethylene and ask what is the value of the integral $\int \phi_1 \phi_2 \, d\tau$. The answer would in fact be *ca.* $\frac{1}{3}$ or $\frac{1}{4}$—$\frac{1}{3}$ if self-consistent-field carbon-atom 2*p*-orbitals are used and $\frac{1}{4}$ if Slater orbitals are employed. So, in adopting this approximation, we really are rather extending our optimistic outlook. Even so, this approximation does not, in practice, appear to make as serious a difference to the calculations as the neglect of magnitudes of this order might lead the reader to think; nevertheless, we shall return to this point in §2.3.

Armed with these approximations, we can now evaluate the Rayleigh Ratio (equation (2-7)), quite easily. Consider first the denominator, $\int \Psi^2 \, d\tau$. From equation (2-1) we can write:

$$\begin{aligned} \int \Psi^2 \, d\tau &= \int (c_1\phi_1 + c_2\phi_2 + \cdots + c_r\phi_r + \cdots + c_n\phi_n)^2 \, d\tau \\ &= \sum_r \int c_r^2 \phi_r^2 \, d\tau + 2\sum_{r<s} \int c_r c_s \phi_r \phi_s \, d\tau \qquad \text{(2-13)} \\ &= \sum_r c_r^2 \int \phi_r^2 \, d\tau + 2\sum_{r<s} c_r c_s \int \phi_r \phi_s \, d\tau \end{aligned}$$

By the normalisation condition (2-11), all the $\int \phi_r^2 \, d\tau$ terms are unity, so that the first term in equation (2-13) is just $(c_1^2 + c_2^2 + \cdots + c_n^2)$; and we have agreed that the integrals $\int \phi_r \phi_s \, d\tau$ are zero because this is the assumption we have just made (equation (2-12)). The second set of terms in (2-13) thus contributes nothing and we conclude that

$$\int \Psi^2 \, d\tau = \sum_r c_r^2 \qquad \text{(2-14)}$$

Now consider the numerator of equation (2-7); we want (again using (2-1)[N6])

$$\int \Psi \mathscr{H} \Psi \, \mathrm{d}\tau = \int (c_1\phi_1 + c_2\phi_2 + \cdots + c_r\phi_r + \cdots + c_n\phi_n)$$

$$\times \mathscr{H}(c_1\phi_1 + c_2\phi_2 + \cdots + c_r\phi_r + \cdots + c_n\phi_n)\mathrm{d}\tau$$

$$= \sum_r \int c_r^2 \phi_r \mathscr{H} \phi_r \, \mathrm{d}\tau + 2\sum_{r<s} \int c_r \phi_r \mathscr{H} c_s \phi_s \, \mathrm{d}\tau$$

$$= \sum_r c_r^2 \int \phi_r \mathscr{H} \phi_r \, \mathrm{d}\tau + 2\sum_{r<s} c_r c_s \int \phi_r \mathscr{H} \phi_s \, \mathrm{d}\tau \qquad (2\text{-}15)$$

We cannot, unfortunately, disregard the second set of terms in (2-15) as we were able to do previously with the overlap integrals, (the corresponding terms in equation (2-13) representing the denominator of the Rayleigh Ratio); they are difficult to handle. The terms involving $\int \phi_i \mathscr{H} \phi_i \, \mathrm{d}\tau$ are really like "little Rayleigh-Ratios" on their own, if one imagined a term $\int \phi_i^2 \, \mathrm{d}\tau$ in the denominator. The expression $\int \phi_i \mathscr{H} \phi_i \, \mathrm{d}\tau / \int \phi_i^2 \, \mathrm{d}\tau$ is thus of the form of an energy; it is in fact the energy one would have if the wave function, in some magic way, had been compelled to be ϕ_i and the problem is determined by a Hamiltonian, $\mathscr{H}$, an effective Hamiltonian we obtain with label i. This is rather like saying what the energy would be if we were to insist upon one of these π-electrons' being on atom i; of course, we cannot in fact insist this and so the quantity does not exist as an observable entity on its own. However, we have now a physical appreciation of what this quantity will be: it is going to be very close to the ionisation potential of the electron on atom i. In fact, the very crudest way of looking at it is to call it the ionisation potential of atom i. It is not *strictly* the ionisation potential because the "$\mathscr{H}$" which is involved here is not just simply the Hamiltonian for an electron on atom i—it is the Hamiltonian (the effective Hamiltonian) which arises when other electrons are present as well. Consequently, what has just been said concerning ionisation potential *is* an approximation. Thus $\int \phi_1 \mathscr{H} \phi_1 \, \mathrm{d}\tau / \int \phi_1 \phi_1 \, \mathrm{d}\tau$ is *approximately equal to the ionisation potential of atom* 1. If the molecule with which we are dealing contains more than one type of atom which participates in the conjugation (say pyridine, in which a nitrogen atom is involved in addition to the carbon atoms) then we shall presumably have to consider different ionisation-potentials for the different types of atoms. Nitrogen is more electronegative than carbon and so numerically the quantity

$$\frac{\int \phi_i \mathscr{H} \phi_i \, \mathrm{d}\tau}{\int \phi_i^2 \, \mathrm{d}\tau}$$

will be larger (negatively) when the ith atom is a nitrogen atom than when it is a carbon atom. The way in which these numerical values are actually estimated will be dealt with later (§3.3); for the moment, we content ourselves by recognising these terms as being an indication of the electronegativity of the atom which is involved at the position labelled i. It is convenient to adopt a matrix notation and write

$$H_{ii} \equiv \int \phi_i \mathscr{H} \phi_i \, \mathrm{d}\tau \tag{2-16}$$

In equation (2-15) we also have cross terms which, in this matrix notation, would be designated H_{ij}, where

$$H_{ij} \equiv \int \phi_i \mathscr{H} \phi_j \, \mathrm{d}\tau \tag{2-17}$$

These terms (with $i \neq j$) have a different interpretation, as will be seen presently; the H_{ij}-terms are in fact a measure of the attraction of the *bond* region, between centres i and j, for electrons, whereas the H_{ii}-terms just dealt with were a measure of the attractiveness of an *atom* (the ith one) for electrons. The H_{ij}-terms therefore are going to tell us something about bonds whereas the H_{ii}-terms will tell us something about atoms.

In this notation, equation (2-15) may be written:

$$\int \Psi \mathscr{H} \Psi \, \mathrm{d}\tau = \sum_r c_r^2 H_{rr} + 2 \sum_{r<s} c_r c_s H_{rs} \tag{2-18}$$

and so, from equations (2-7), (2-14), (2-15) and (2-18), the complete Rayleigh-Ratio may be expressed as

$$y = \frac{\sum_r c_r^2 H_{rr} + 2 \sum_{r<s} c_r c_s H_{rs}}{\sum_r c_r^2} \tag{2-19}$$

It is this expression which we now require to make stationary, with respect to the $\{c_r\}$. Let us first consider making y stationary with respect to variation of c_1—*i.e.* consider making $\partial y/\partial c_1 = 0$. First, we must cross-multiply in order to write equation (2-19) as

$$\begin{aligned} (c_1^2 + c_2^2 + \cdots + c_r^2 + \cdots + c_n^2)\partial y \\ = (c_1^2 H_{11} + c_2^2 H_{22} + \cdots + c_r^2 H_{rr} + \cdots + c_n^2 H_{nn}) \\ + 2(c_1 c_2 H_{12} + c_1 c_3 H_{13} + \cdots + c_{n-1} c_n H_{n-1,n}) \end{aligned} \tag{2-20}$$

Next, differentiating partially with respect to c_1 gives

$$\begin{aligned} 2c_1 y \partial c_1 + (c_1^2 + c_2^2 + \cdots + c_r^2 + \cdots + c_n^2) y \\ = 2c_1 H_{11} \partial c_1 + 2(c_2 H_{12} + c_3 H_{13} + \cdots + c_r H_{1r} + \cdots + c_n H_{1n}) \partial c_1 \end{aligned} \tag{2-21}$$

i.e.,

$$\frac{\partial y}{\partial c_1} \equiv 0 = \frac{2\{c_1(H_{11} - y) + c_2 H_{12} + c_3 H_{13} + \cdots + c_r H_{1r} + \cdots + c_n H_{1n}\}}{(c_1^2 + c_2^2 + \cdots + c_r^2 + \cdots + c_n^2)} \tag{2-22}$$

Since the denominator of equation (2-22) is non-zero and finite it is the numerator which must vanish in order for the condition $\partial y/\partial c_1 = 0$ to be satisfied. This, therefore, leads to

$$c_1(H_{11} - y) + c_2 H_{12} + c_3 H_{13} + \cdots + c_r H_{1r} + \cdots + c_n H_{1n} = 0 \tag{2-23}$$

It is this equation which must be satisfied in order that the Rayleigh Ratio (equation (2-7)) shall be stationary for variations, specifically, of c_1. Similarly, differentiating y in equation (2-19) partially with respect to c_2 leads to

$$c_1 H_{21} + c_2(H_{22} - y) + c_3 H_{23} + \cdots + c_r H_{2r} + \cdots + c_n H_{2n} = 0 \tag{2-24}$$

In general, differentiating y partially with respect to c_r yields

$$c_1 H_{r1} + c_2 H_{r2} + \cdots + c_r(H_{rr} - y) + \cdots + c_n H_{rn} = 0. \tag{2-25}$$

Thus, when y has been differentiated partially with respect to each c_i in turn, $i = 1, 2, \ldots, r, \ldots, n$, a set of n homogeneous equations is obtained which are called the *secular equations* of the system. They are of the general form:

$$\left.\begin{array}{l}
c_1(H_{11} - y) + c_2 H_{12} + \cdots + c_r H_{1r} + \cdots + c_n H_{1n} = 0 \\
c_1 H_{21} + c_2(H_{22} - y) + \cdots + c_r H_{2r} + \cdots + c_n H_{2n} = 0 \\
\vdots \\
c_1 H_{r1} + c_2 H_{r2} + \cdots + c_r(H_{rr} - y) + \cdots + c_n H_{rn} = 0 \\
\vdots \\
c_1 H_{n1} + c_2 H_{n2} + \cdots + c_r H_{nr} + \cdots + c_n(H_{nn} - y) = 0
\end{array}\right\} \tag{2-26}$$

It is these homogeneous equations which must be solved in order to obtain the $\{c_r\}$ we require and, at the same time, the corresponding values of y which these sets of coefficients cause to be stationary. These values of y, the roots of the secular equations (2-26), will be taken to represent the *energy* of the corresponding molecular-orbital. For this reason, y will henceforth be replaced by the symbol, ε, to denote an energy. At the same time, we can write the secular equations in a more compact way, as follows:

$$(H_{rr} - \varepsilon)c_r + \sum_s{}' H_{rs} c_s = 0, \qquad r = 1, 2, \ldots, n \tag{2-27}$$

where a value of ε which will satisfy this equation is the energy of a molecular orbital and the summation, $\sum_s'$, is over the $s \neq r$, the prime indicating omission of the term $s = r$ from the summation.

Now equations (2-26) and (2-27) are a set of n homogeneous equations in n unknowns—the latter being one set of $(c_1, c_2, \ldots, c_n)$ which make the Rayleigh Ratio stationary. The equations are called "homogeneous" since, in each of the n equations, zero occurs on the right-hand side. In fact, it is not immediately obvious that this type of equation does have solutions—there are in a way too many equations for the effective number of unknowns. It is appropriate to say "effective" number because what we have at this stage, from these equations, are really only the *ratios* of the coefficients. This is immediately evident because one can multiply each c_r by some constant (3, say) and the equations will still be satisfied. So, contrary to initial appearances, there really are only $(n - 1)$ independent unknowns in equations (2-26)—and yet there are n equations. One set of trivial solutions which would clearly satisfy (2-26) would be $c_1 = c_2 = c_3 = \cdots = c_r = \cdots = c_n = 0$. In order, however, to obtain non-trivial solutions to a set of n homogeneous equations of this kind, we must have the situation in which there are really only $(n - 1)$ *independent* equations. This can be brought about if one of the equations involving $(c_1, c_2, \ldots, c_n)$ of (2-26) is a linear combination of the others—*i.e.* can be obtained by addition and subtraction of appropriate multiples of the other $(n - 1)$ equations. The theory of determinants and homogeneous equations establishes that this will be the case if the determinant of the coefficients of the variables,[N7] $\{c_r\}$, in equations (2-26) (called the *secular determinant*) vanishes, *i.e.*,

$$\begin{vmatrix} H_{11} - \varepsilon & H_{12} & \cdots & H_{1r} & \cdots & H_{1n} \\ H_{21} & H_{22} - \varepsilon & \cdots & H_{2r} & \cdots & H_{2n} \\ \vdots & \vdots & & \vdots & & \vdots \\ H_{r1} & H_{r2} & \cdots & H_{rr} - \varepsilon & \cdots & H_{rn} \\ \vdots & \vdots & & \vdots & & \vdots \\ H_{n1} & H_{n2} & \cdots & H_{nr} & \cdots & H_{nn} - \varepsilon \end{vmatrix} = 0 \qquad (2\text{-}28)$$

This is the condition which must be satisfied in order that the secular equations ((2-26) and (2-27)) shall have non-trivial solutions. Now, in condition (2-28), the only quantity which is not known (or, at least, the only quantity which is *presumed* not to be known) is the energy, ε (see note 7); it occurs on the diagonal—and, in this approximation, *only* on the diagonal. On expansion, this determinant leads to an nth-order polynomial in ε, of the form

$$(-1)^n \varepsilon^n + a_1 \varepsilon^{n-1} + a_2 \varepsilon^{n-2} + \cdots + a_n = 0 \qquad (2\text{-}29)$$

Put in matrix notation, in which $\mathbb{H} \equiv [H_{ij}]$, equation (2-28) may be written

$$|\mathbb{H} - \varepsilon \mathbb{1}_{n \times n}| = 0 \tag{2-30}$$

(where $\mathbb{1}_{n \times n}$ is the unit matrix of dimension $n \times n$). When equation (2-28) is written in this form, equation (2-29) is seen to be the characteristic polynomial of the matrix, $\mathbb{H}$; the n roots of this polynomial are therefore the eigenvalues of the matrix, $\mathbb{H}$. Because of the Hermitian nature of the operator, $\mathscr{H}$, referred to earlier, the matrix, $\mathbb{H}$, is an Hermitian matrix (specifically, it is a real-symmetric matrix) and consequently all its eigenvalues (the roots, $\{\varepsilon_I\}$, of (2-29)) are real. This is, of course, a necessary requirement because we are going to say that the (at-most-n-distinct) roots of (2-29), $\{\varepsilon_I\}$, $I = 1, 2, \ldots, n$, represent the *energies* of the n molecular-orbitals of the system. This set of n values of ε, $\{\varepsilon_I\}$, $I = 1, 2, \ldots, n$, are the particular ones which, when inserted into (2-28), make the determinant on the left-hand side of this equation zero, and thus ensure that the homogeneous equations, (2-26), have non-trivial solutions. So there are n *molecular*-oribtals associated with the conjugated system in question. We started with n *atomic*-orbitals and have ended up with n *molecular*-orbitals. This is a completely general rule in minimal basis-set calculations of this sort.

We now examine how, in principle, given one of the roots of (2-29) (consider the Ith—call it ε_I) which makes the secular determinant in (2-28) vanish, we can obtain the weighting-coefficients, $\{c_{Ir}\}$, $r = 1, 2, \ldots, n$, which satisfy the set of homogeneous equations (2-27) *for this particular value of* ε. It will be recalled that in the discussion following equation (2-1) in the section concerning the LCAO-approach (§2.1), it was stated that "Ψ" in equation (2-1) should strictly be labelled with a subscript as "Ψ_I" since there is always more than one combination of the given basis atomic orbitals which will form what, on the Variation-Principle's criterion, we may regard as a "suitable" molecular orbital. In fact, our analysis here has confirmed that there are in total n such MO's, $\{\psi_I\}$, $I = 1, 2, \ldots, n$, the energies of which are the n roots of the characteristic polynomial (equation (2-29)) of the matrix, $\mathbb{H}$—*i.e.*, of the secular determinant (equation (2-28)). "c_r" in equation (2-1) should therefore be replaced by "c_{Ir}" (meaning "the weighting-coefficient of the atomic orbital centred on atom r in the Ith LCAO-MO") and this additional subscript, omitted from equation (2-1), should be reinstated so that this latter equation reads:

$$\Psi_I = \sum_r c_{rI} \phi_r \tag{2-31}$$

and the n secular-equations, (2-27), which ε_I and $\{c_{Ir}\}$ satisfy can be written:

$$(H_{rr} - \varepsilon_I) c_{Ir} + \sum_s{}' H_{rs} c_{Is} = 0, \qquad r = 1, 2, \ldots, n \tag{2-32}$$

We have agreed that ε_I is such that when ε in the secular determinant (equation (2-28)) is assigned the value ε_I, the determinant vanishes—this was by arrangement, for ε_I is a root of (2-29). If the determinant of the coefficients of the $\{c_{Ir}\}$ in the n secular-equations (2-32) is zero, then from the theory of determinants this means that one row of the determinant is a linear combination of the others—in other words, it means that one of the n equations in (2-32) is redundant. Contrary to initial appearances, therefore, there are only $(n - 1)$ *independent*, simultaneous equations in (2.32), but there are n unknowns, $\{c_{Ir}\}$, $r = 1, 2, \ldots, n$. Consequently, all that we can find by elimination from these equations is the *ratio*[N8] of the weighting coefficients, c_{I1}: c_{I2}: $c_{I3} \cdots c_{In}$, associated with the Ith MO, and not their absolute values. One further condition is required if we are to determine the $\{c_{Ir}\}$, $r = 1, 2, \ldots, n$, absolutely. For this condition we make appeal to the physically-sensible requirement that Ψ_I be normalised—*i.e.* that

$$\int \Psi_I^2 \, \mathrm{d}\tau = 1 \tag{2-33}$$

From equation (2-14), in the neglect-of-overlap case, this implies

$$c_{I1}^2 + c_{I2}^2 + \cdots + c_{Ir}^2 + \cdots + c_{In}^2 = \sum_r c_{Ir}^2 = 1 \tag{2-34}$$

Condition (2-34), therefore, together with the ratio, c_{I1}:c_{I2}:$c_{I3} \cdots c_{In}$, determined from equations (2-32), is thus in principle sufficient to establish the *absolute* magnitudes of $\{c_{Ir}\}$, $r = 1, 2, \ldots, n$.

For an arbitrary, π-electron system made up from n basis atomic-orbitals, we have now, in principle, described how to find, by the Variation Method, the energies, $\{\varepsilon_I\}$, $I = 1, 2, \ldots, n$, and the LCAO-combinatorial weighting-coefficients, $\{c_{Ir}\}$, $I = 1, 2, \ldots, n$, $r = 1, 2, \ldots, n$, of the n LCAO-molecular-orbitals, Ψ_I, $I = 1, 2, \ldots, n$, which describe the system.

2.3 A Word on Inclusion of Overlap

It was stated earlier that in the simplest assumption we were going to make we would assume that the basis atomic-orbitals were normalised, *i.e.*,

$$S_{rr} = \int \phi_r \phi_r \, \mathrm{d}\tau = 1 \tag{2-35}$$

and orthogonal, *i.e.*,

$$S_{rs} = \int \phi_r \phi_s \, \mathrm{d}\tau = 0 \qquad (r \neq s) \tag{2-36}$$

(These two statements may conveniently be combined by use of the "Kronecker-δ" symbol, δ_{ij}, which is equal to 1 when the suffices are the same $[i = j]$ and zero when they are different $[i \neq j]$ *i.e.*,

$$S_{rs} = \delta_{rs} \tag{2-37}$$)

In reality, as we have seen, the basis atomic orbitals are *not* orthogonal; S_{rs} for adjacent carbon-$2p_\pi$ is of the order of 0.3—which is very far from being zero. One might therefore fear that all the analysis so-far outlined would be thrown out completely if we were to include these overlap terms which have previously been neglected.

The complete details of including overlap will not be given here for it is necessary to follow much the same process including overlap as we have done excluding it. The Rayleigh Ratio will still have to be dealt with and it is as before (equation (2-7)); the numerator of the Rayleigh Ratio will be as before (equations (2-15) and (2-18)); however, if we do not neglect the overlap terms, the denominator of the Rayleigh Ratio must be given by the full form of equation (2-13) and not the reduced form of equation (2-14)—*i.e.*, the denominator is

$$\int \Psi^2 \, d\tau = \sum_r c_r^2 + 2 \sum_{r<s} c_r c_s S_{rs} \tag{2-38}$$

(from which, for simplicity, the MO-subscript, I, has again been omitted).

The Rayleigh Ratio thus becomes, instead of (2-19),

$$y = \frac{\sum_r c_r^2 H_{rr} + 2 \sum_{r<s} c_r c_s H_{rs}}{\sum_r c_r^2 + 2 \sum_{r<s} c_r c_s S_{rs}} \tag{2-39}$$

By going through the process of finding $\partial y/\partial c_r$, $r = 1, 2, \ldots, n$, in a way exactly analogous to that represented by equations (2-19)–(2-27) and again replacing y by ε, we can obtain the following set of secular equations, analogous to equation (2-27) in the neglect-of-overlap case:

$$(H_{rr} - \varepsilon)c_r + \sum_s{}' (H_{rs} - \varepsilon S_{rs})c_s = 0 \qquad r = 1, 2, \ldots, n. \tag{2-40}$$

The zero secular-determinant equation is then (*cf.* equation (2-28))

$$\begin{vmatrix} H_{11} - \varepsilon & H_{12} - \varepsilon S_{12} & \cdots & H_{1r} - \varepsilon S_{1r} & \cdots & H_{1n} - \varepsilon S_{1n} \\ H_{21} - \varepsilon S_{21} & H_{22} - \varepsilon & \cdots & H_{2r} - \varepsilon S_{2r} & \cdots & H_{2n} - \varepsilon S_{2n} \\ \vdots & \vdots & & \vdots & & \vdots \\ H_{r1} - \varepsilon S_{r1} & H_{r2} - \varepsilon S_{r2} & \cdots & H_{rr} - \varepsilon & \cdots & H_{rn} - \varepsilon S_{rn} \\ \vdots & \vdots & & \vdots & & \vdots \\ H_{n1} - \varepsilon S_{n1} & H_{n2} - \varepsilon S_{n2} & \cdots & H_{nr} - \varepsilon S_{nr} & \cdots & H_{nn} - \varepsilon \end{vmatrix}$$

or, in more-compact matrix-notation (*cf.* equation (2-30)),

$$|\mathbb{H} - \varepsilon\mathbb{S}| = 0 \tag{2-41}$$

where $\mathbb{S}$ is the $n \times n$ overlap-matrix, equal to $[S_{ij}]$.

Expansion of equation (2-41) again leads to an nth-order polynomial in ε, the roots of which may again be interpreted as being the energies of the several (n) molecular-orbitals of the system. So there really is not much trouble, in principle, in including the overlap integral. Historically, people have tended not to include it because it turns out not to make much numerical difference; and since the object of the approach we are discussing is to have a simple picture and not a really elaborate one we have to ask ourselves whether it really is worthwhile making refinements. However nice they might be and, in a sense, satisfying to our *ego*, such refinements are not worthwhile making if we are still left with very doubtful assumptions—and we still *are* left with very doubtful assumptions, some of which will emerge as we go on! So although, in principle, one can deal with the secular determinant in the way in which it is presented in equation (2-41), we shall mostly neglect $\mathbb{S}$ and handle the secular determinant in its form given in equation (2-28).

2.4 A More-Rigorous Demonstration of σ-π Separability

The second comment which it is appropriate to make here is that we can already see the reason for σ-π separation by this method. In the LCAO-scheme we represent a molecular orbital as a linear combination of atomic orbitals, as *per* equation (2-1). We are going now to ask what kind of conclusion would be reached if, in the summation of equation (2-1), we supposed that some of the ϕ_i's were σ-orbitals, in addition to the orbitals of π-symmetry already considered. Of course, these σ-orbitals need not necessarily be included—it behoves us to *show* that this is so—but let us for the moment (again, for simplicity, dropping the MO index-number) write

$$\Psi = \sum_{r=1}^{m} c_r \phi_r \tag{2-42}$$

and suppose that atomic basis-orbitals 1 to n are of π-type, and orbitals $(n + 1)$ up to m are of σ-type. As an example, Fig. 2-1 shows a $2p_\sigma$- orbital, relative to the plane in which the nuclei which comprise the conjugated system are located, and a $2p_\pi$-orbital, relative to that plane.

In principle, we can follow exactly the same procedure as before: that is to say, we can work out what the Rayleigh ratio is, associated with the wave function (equation (2-7)), obtain y as a function of the $\{c_r\}$, then find $\partial y/\partial c_r$

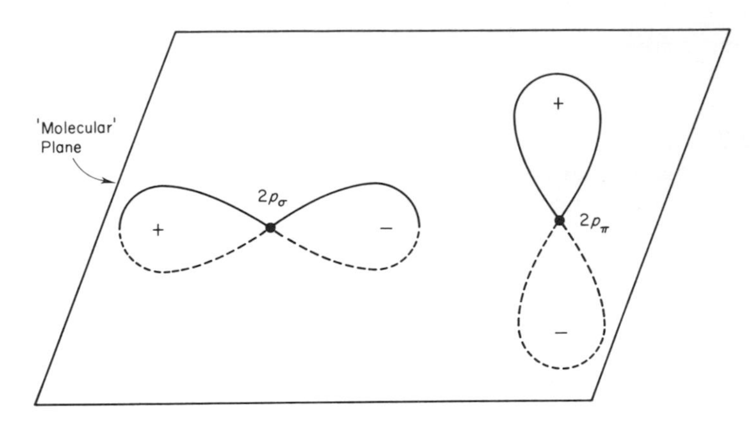

FIG. 2-1 Orbitals of σ- and π-symmetry, with respect to a molecular plane.

for each c_r, $r = 1, 2, \ldots, n$, etc. Mathematically, it is precisely the same. Formally the same set of secular equations as those in equation (2-26) will be obtained if overlap is neglected, and as equation (2-41) if it is not. By eliminating the $\{c_r\}$ we obtain a secular determinant which formally is identical with what we had before (equation (2-28)). It is reproduced again below, for we are going to partition it, since this is where the important conclusion of the present argument about σ-π separation comes. The secular determinant arising from this treatment may be partitioned into those terms (forming an $n \times n$ sub-matrix) which come from π and those (forming an $(m\text{-}n) \times (m\text{-}n)$ block) which arise from the σ-basis orbitals in (2-42). *i.e.*,

$$
\begin{array}{cc}
 & \begin{array}{cc} \overset{\pi}{\longleftarrow n \longrightarrow} & \overset{\sigma}{\longleftarrow m-n \longrightarrow} \end{array} \\
\begin{array}{c} \pi \; n \\ \\ \sigma \; m-n \end{array} &
\left|\begin{array}{cccc|cccc}
H_{11}-\varepsilon & H_{12} & \cdots & H_{1n} & & & & \\
H_{21} & H_{22}-\varepsilon & \cdots & H_{2n} & & & & \\
\vdots & \vdots & & \vdots & & & & \\
H_{n1} & H_{n2} & \cdots & H_{nn}-\varepsilon & & & & \\
\hline
 & & & & H_{n+1,n+1}-\varepsilon & & \cdots & H_{n+1,m} \\
 & & & & H_{n+2,n+1} & H_{n+2,n+2}-\varepsilon & \cdots & H_{n+1,m} \\
 & & & & & \vdots & & \vdots \\
 & & & & H_{m,n+1} & H_{m,n+2} & \cdots & H_{mm}-\varepsilon
\end{array}\right| = 0
\end{array}
\tag{2.43}
$$

The first n terms arise from π-atomic-orbitals so that the upper-left $n \times n$ sub-matrix represents π-π interactions; the lower-right $(m - n) \times (m - n)$ sub-matrix similarly represents σ-σ interactions.

What we wish to show is that the upper-right ($n \times (m - n)$) sub-matrix comprises nothing but *zero* entries and that the lower-left ($(m - n) \times n$) sub-matrix is also a zero matrix. If we can show this then it will mean that the secular determinant really does factorise into two entirely separate ones. On the one hand there is the upper-left block which is precisely the familiar π-description which we have been dealing with in most of the discussion so far, and on the other hand there is the bottom-right block which would be a σ-description. Further, in this situation, we would be dealing *either* with π-electrons governed, as before, by the kind of discussion we have given *or*, alternatively, with the σ-electrons which are separated completely from the π, with no interaction between the two. So, in order to show that σ and π do not mix we must simply show that all the elements in the upper-right and lower-left corners of the secular determinant (equation (2-43)) are zero. This is not at all difficult and we proceed to show it on the grounds of symmetry. This in itself should not be surprising since, as we have seen, "σ" and "π" are, in reality, symmetry labels.

Let us take as an example a p_σ-orbital centred on atom r and a p_π-orbital on atom s (Fig. 2-2) and keep in mind the plane of the nuclei of the con-

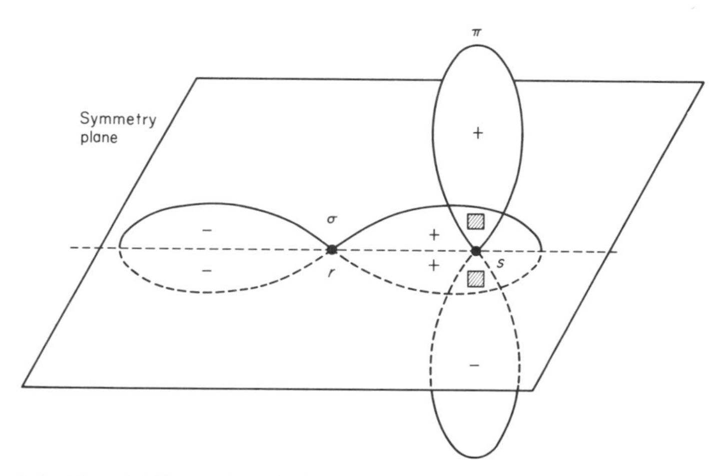

FIG. 2-2 Pictorial illustration of (algebraic) zero-overlap between σ- and π-orbitals.

jugated system in which these interactions are taking place, and with respect to which the designations of the orbitals concerned as "σ" or "π" are referred. The p_π-orbital can be considered to have its "axis" perpendicular to the molecular plane while the p_σ-orbital has its axis in the plane. We have already observed that a π-orbital is an odd function with respect to reflection in the molecular plane while the σ-orbital is an even function with respect to this operation. The π-orbital is thus shown in Fig. 2-2 as being positive above the plane and negative below the plane, while that part of the p_σ-function which overlaps with it is shown as positive, both above and below the plane. Consider first an overlap integral

$$\int \phi_r^\sigma \phi_s^\pi \, d\tau \tag{2-44}$$

with orbitals centred on r and s being σ and π, respectively, as in Fig. 2-2. Such an overlap integral is really the sum of individual contributions which come from every little bonding-element in the region of overlap. Let us consider two such bonding-elements, one above the plane and the other its exact mirror-image below (Fig. 2-2); in these two elements, since they are mirror images, ϕ_r^σ has the same value and ϕ_s^π has changed its sign, but is otherwise numerically the same. Consequently, the product $\phi_r^\sigma \phi_s^\pi$ has equal value but opposite sign in these two regions and so, if we were to add up the contributions which come from the integral in those two regions, they would cancel. What is more, we can cancel the contribution of every other element of volume from above the plane with a corresponding element below the plane and so conclude that

$$S_{rs} \equiv \int \phi_r^\sigma \phi_s^\pi \, d\tau \equiv 0 \tag{2-45}$$

by symmetry.[N9]

Let us now consider what would happen with the corresponding matrix-elements of the effective Hamiltonian,

$$H_{rs} = \int \phi_r^\sigma \mathscr{H} \phi_s^\pi \, d\tau \tag{2-46}$$

Consider, in particular, the Hamiltonian; it has a kinetic-energy part and a potential-energy part, the expressions for which are given in equations (2-4) and (2-5). The kinetic-energy operator is $(h^2/8\pi m)\nabla^2$. Now ∇^2 (which is explicitly defined in equation (2-4)) is unchanged when x is replaced by $(-x)$, y is replaced by $(-y)$ and/or z is replaced by $(-z)$, for it contains only terms (such as $\partial^2/\partial x^2$) which are *second order* in these variables—that is to say, ∇^2 is an *even* operator in x, y and z. In particular, if we arbitrarily consider the molecular plane to be the xy-plane, then we are interested in the behaviour of ∇^2 with respect to the change, $z \to (-z)$; as we have argued above, ∇^2 is indeed even with respect to this replacement and is thus even with respect to reflection in the molecular plane. By symmetry, the potential-energy-operator part of the effective Hamiltonian, $\mathscr{V}_{\text{effective}}$, of equation (2-5) must also be the same above and below the plane. Consequently, the entire effective-Hamiltonian, $\mathscr{H}$, is even with respect to reflection in the molecular plane and so the arguments used previously concerning overlap integrals between σ- and π-atomic-orbitals carries through to the matrix elements, H_{rs}, and we can say that

$$H_{rs} \equiv \int \phi_r^\sigma \mathscr{H} \phi_s^\pi \, d\tau \equiv 0, \tag{2-47}$$

identically, by symmetry.[N10]

These arguments, therefore, mean that all the elements which come into our hypothetical secular-determinant mixing σ and π (equation (2-43)) are zero—*i.e.*, the $(n \times (m - n))$ sub-matrix in the upper-right corner of (2-43) and the $((m - n) \times n)$ sub-matrix in the lower-left corner of (2-43) are both *zero* matrices. We have thus made our case that one cannot mix in any σ-orbitals with π-orbitals; hence we speak of this *π-σ separability*. However, the reader should be cautioned that the word "separability" is an extremely delicate one. It does not mean that the π-electrons are, as-it-were, "unaware" of the "existence" of the σ-electrons; it certainly does not mean that. The two types of electrons *do* influence each other since, for example, $\mathscr{V}_{\text{effective}}$ in the effective Hamiltonian (equation (2-5)) will include interactions with the σ-electrons; there will be Coulomb-interactions and these two types of electrons will repel each other. So it is not true to say that the π-electrons are totally and blissfully unaware of the presence of the σ-electrons! It does mean, however, that when we write out an LCAO molecular-orbital to describe them, we do not need to include, in the same molecular orbital, both σ- *and* π atomic-orbitals.

That argument depended on the symmetry condition outlined above—in fact, all the discussion in this subsection is fundamentally one of symmetry. If that symmetry rule is broken, then the σ-π-separability argument will no longer hold. The following few examples will illustrate the meaning of this last statement.

Example 1

Consider a π-complex between a carbon-carbon "double"-bond and an Ag^+-ion, as illustrated in Fig. 2-3.

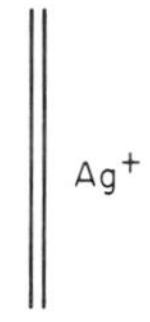

FIG. 2-3 π-complex between a carbon-carbon double-bond and an Ag^+ ion.

Here, the presence of the Ag^+-ion disturbs the symmetry of the electron-distribution above and below the plane of the conjugated system containing the double bond in question. One of the vital conditions for σ-π separability is thus spoiled.

Example 2

The next example in which σ-π separability breaks down concerns that amusing class of molecules, the cyclophanes (Fig. 2-4). It has been customary,

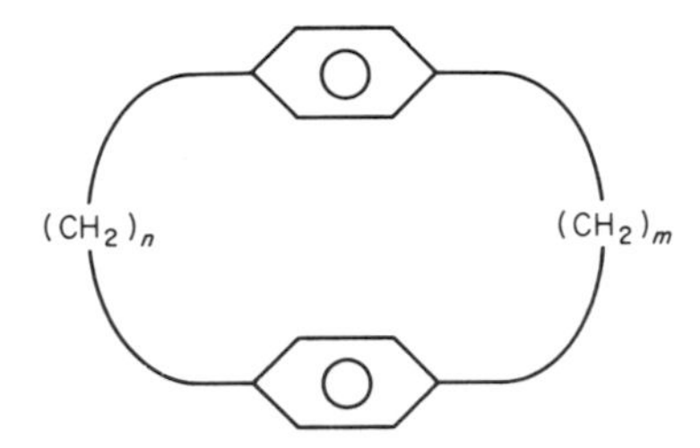

FIG. 2-4 An (n, m)-para-cyclophane molecule.

when dealing with systems like these, to treat the two benzenoid groups initially as isolated, planar π-electron-systems, to talk about separate π-molecular-orbitals for each of the benzenoid rings—and then perhaps to combine the lower lobes of the π-orbitals on the upper ring with upper lobes of the π-orbitals on the lower ring, according to symmetry rules. This is not a strictly correct procedure if only because the benzenoid rings themselves are by no means rigorously planar; in fact, they are frequently "boat-shaped", as can be shown by X-ray crystallography. As an example, Fig. 2-5 illustrates schematically a perspective view of the cyclophane in which $m = n$ (of Fig. 2-4) $= 2$. Since the benzenoid rings are no longer planar, we cannot speak of a "molecular plane" and so we cannot even make the σ-π *distinction*, let alone talk about a σ-π separability.

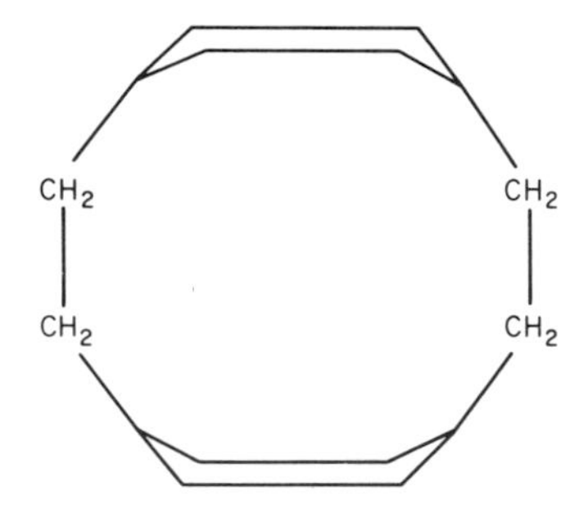

FIG. 2-5 2,2-para-cyclophane—a perspective view.

This latter example in particular should therefore be heeded as a warning that we really are supposing the systems we are discussing to be planar when we separate a molecular orbital into σ- and π-parts.

2.5 The α- and β-Notation

We now revise our notation a little in order to bring it into line with what has become traditional in Hückel theory; it is called the *α, β-notation.*

The matrix components of some effective-Hamiltonian (represented in equation (2-5)) were quite properly given, in the earlier discussion, the normal symbols for the matrix components of a Hamiltonian—namely H_{rr} and H_{rs}, defined in equations (2-16) and (2-17). However, quite early on in the development of the subject, Hückel and others gave the following different symbols to them: H_{rr} was referred to as α_r and known as the *Coulomb integral for atom r*; H_{rs} was referred to as β_{rs} and known as the *resonance integral for the bond between atoms r and s.* As argued earlier, α_r is a measure of the electron-attracting power of the atom r and can be considered, in a crude way, to represent the electronegativity of that atom. Similarly, β_{rs} can be interpreted as the electron-attracting power of the bond $r - s$, as distinct from the individual atoms, r and s, which are involved in it.

In this notation, the Rayleigh Ratio (energy) corresponding to a given set of coefficients, is

$$\varepsilon = \frac{\sum_r c_r^2 \alpha_r + 2 \sum_{r<s} c_r c_s \beta_{rs}}{\sum_r c_r^2} \tag{2-48}$$

where, in general, the second summation in the numerator involves pairs of atoms, r and s, which are not bonded, as well as those which are. The energy of an electron in a molecular orbital is thus related to the electron-attracting power of the several individual atoms in the molecule (α_r) and to that of the various atom-pairs it comprises (β_{rs}). For atoms which are able to form strong double-bonds, for example, the resonance integral will be large, numerically; otherwise it is quite small—in particular, the resonance integral, β_{rs}, turns out to be quite small *except when atoms r and s are already connected by a σ-bond*. Recall that β_{rs} is the integral of the product of the atomic orbital on atom r with the result of the effective Hamiltonian's acting on the atomic orbital centred on atom s—*i.e.*,

$$\beta_{rs} = \int \phi_r \mathscr{H} \phi_s \, d\tau \tag{2-49}$$

Now an atomic orbital on atom r naturally has a substantial numerical value only in the neighbourhood of atom r. In (2-49) we have to consider the product of this function with $\mathscr{H}\phi_s$; even after being operated on by $\mathscr{H}$, ϕ_s still has its largest value in the vicinity of atom s. Consequently, if atoms r and s are well-separated there will be very little overlap between ϕ_r and $\mathscr{H}\phi_s$ and thus there will be effectively no part of space in which the integrand $\phi_r\mathscr{H}\phi_s$ is measurable; we therefore expect the integral in equation (2-49) to be zero, or nearly so. The resonance integral, β_{rs}, is thus zero, or fairly close to zero, except for bonded neighbours, r and s. Of course, neglect of non-bonded resonance-integrals is certainly not mandatory in this approach —one may include β's between non-nearest neighbours but it is found that such an inclusion does not in practice make very much difference, numerically, to the results of the calculations. The actual estimation of appropriate values for the $\{\alpha_r\}$- and $\{\beta_{rs}\}$-parameters will be considered presently (in Chapter Three).

In the α, β-notation, the secular determinant (without overlap—equation (2-28)) is as follows:

$$\begin{vmatrix} \alpha_1-\varepsilon & \beta_{12} & \beta_{13} & \cdots & \beta_{1r} & \cdots & \beta_{1n} \\ \beta_{21} & \alpha_2-\varepsilon & \beta_{23} & \cdots & \beta_{2r} & \cdots & \beta_{2n} \\ \vdots & \vdots & \vdots & & \vdots & & \vdots \\ \beta_{r1} & \beta_{r2} & \beta_{r3} & \cdots & \alpha_r-\varepsilon & \cdots & \beta_{rn} \\ \vdots & \vdots & \vdots & & \vdots & & \vdots \\ \beta_{n1} & \beta_{n2} & \beta_{n3} & \cdots & \beta_{nr} & \cdots & \alpha_n-\varepsilon \end{vmatrix} = 0 \qquad (2\text{-}50)$$

and the rth secular equation (equation (2-27)) is

$$(\alpha_r-\varepsilon)c_r + \sum_s{}' \beta_{rs}c_s = 0 \qquad (2\text{-}51)$$

2.6 The Hückel Assumption for Hydrocarbons

In the case of conjugated hydrocarbons, Hückel made the following assumptions and approximations, now known as the Hückel approximations.

1) The Coulomb integrals, α_r, are set equal to the common value, α, for all carbon atoms, irrespective of their environment in the conjugated system.

$$\alpha_r = \alpha, \qquad r = 1, 2, 3, \ldots, n \qquad (2\text{-}52)$$

2) All resonance integrals, β_{rs}, are taken to be zero if atoms r and s are not bonded neighbours. This is sensible by virtue of the arguments advanced in the previous section (§2·5). Furthermore, if r and s *are* bonded neighbours, then β_{rs} is given the common value, β, for all carbon-carbon bonds,

irrespective of their environment in the conjugated system, *i.e.*

$$\begin{aligned}\beta_{rs} &= \beta,\ r, s \text{ bonded} \\ &= 0, \text{ otherwise, for all } r \neq s \\ &\qquad r = 1, 2, 3, \ldots, n \\ &\qquad s = 1, 2, 3, \ldots, n\end{aligned} \tag{2-53}$$

2.7 A Worked Example: Butadiene

We now work through in detail an example of the application of the Hückel method to a specific molecule; in its simplest form, the calculation will be characterised by just two empirical parameters, α and β. Butadiene is selected as the example and the molecular-orbital energies and LCAO combinatorial coefficients calculated for it here will be used in later discussions of other quantities derivable from them (such as charge, bond order and free valence discussed in Chapter Four). Butadiene has four carbon atoms, the σ-bond connectivity of which may conveniently be depicted schematically as in Fig. 2-6.

Notice that Fig. 2-6 illustrates a very important aspect of Hückel theory which it is as well to emphasise here; namely that although a molecule like

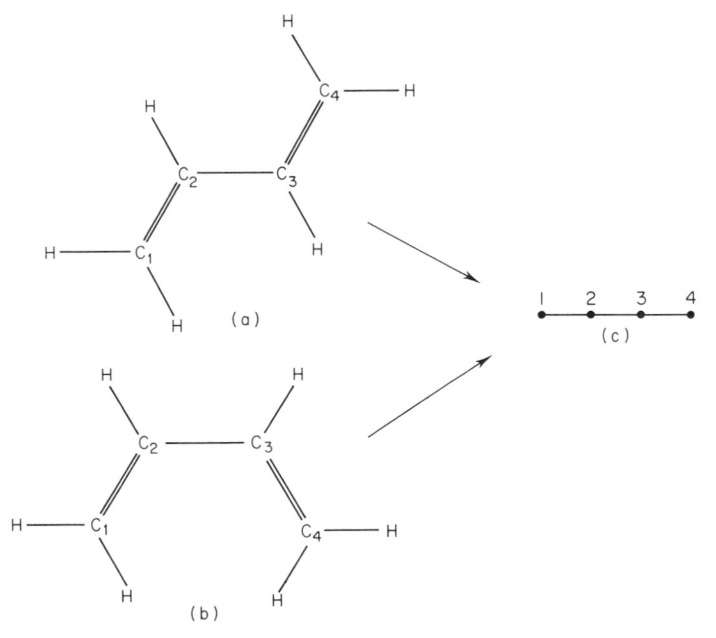

FIG. 2-6 Reduction of both *cis*- and *trans*-butadiene to their equivalent graph which illustrates the carbon-carbon σ-bond connectivity of *both* forms of butadiene.

butadiene is known to exist in the *cis*- and *trans*-forms as shown in Fig. 2-6(a) and (b), both are schematically represented for the purposes of a Hückel calculation as in Fig. 2-6(c). That is, Hückel theory is concerned *only with the σ-bond connectivity of a conjugated system and not with molecular geometry*[N11]. In a more subtle way, Hückel theory does have something to say about molecular geometry in so far as it may be used (*via* the concept of bond *order*) to predict bond *lengths*—but *not* bond *angles*. However, these are points which we shall return to later (Chapter Four); for the moment, we continue our application of the simple Hückel-method to butadiene.

We may immediately construct the Hückel Hamiltonian-matrix for butadiene; it is going to be 4×4. The Coulomb integral, α, will occur along the diagonal since all atoms in the conjugated system are carbon and the only resonance integrals, β, which are non-zero are those between centres which are bonded. Thus the (1-2)-, (2-3)-, and (3-4)-elements are β, and so also, of course, are the (2-1)-, (3-2)-, and (4-3)-elements—the matrix must clearly be symmetrical since, if atom i is joined to atom j, then certainly atom j is joined to atom i! Thus we have, with the vertex labelling adopted in Fig. 2-6,

$$\mathbb{H} = \begin{pmatrix} \alpha & \beta & 0 & 0 \\ \beta & \alpha & \beta & 0 \\ 0 & \beta & \alpha & \beta \\ 0 & 0 & \beta & \alpha \end{pmatrix} \tag{2-54}$$

In the zero-overlap case, the secular determinant is $|\mathbb{H} - \varepsilon \mathbb{1}_{4\times 4}|$ and so we have

$$\begin{vmatrix} \alpha - \varepsilon & \beta & 0 & 0 \\ \beta & \alpha - \varepsilon & \beta & 0 \\ 0 & \beta & \alpha - \varepsilon & \beta \\ 0 & 0 & \beta & \alpha - \varepsilon \end{vmatrix} = 0 \tag{2-55}$$

At this point it is a great convenience to divide each element of the determinant by β (since we are dealing with a zero determinant, this is perfectly permissible) to obtain

$$\begin{vmatrix} \frac{\alpha - \varepsilon}{\beta} & 1 & 0 & 0 \\ 1 & \frac{\alpha - \varepsilon}{\beta} & 1 & 0 \\ 0 & 1 & \frac{\alpha - \varepsilon}{\beta} & 1 \\ 0 & 0 & 1 & \frac{\alpha - \varepsilon}{\beta} \end{vmatrix} = 0 \tag{2-56}$$

and to measure energy with *reference* to α (*i.e.*, to take the Coulomb integral, α, as the energy zero) and in *units* of β, *i.e.*, to set

$$x = \frac{\varepsilon - \alpha}{\beta}. \tag{2-57}$$

The secular determinant for butadiene then becomes

$$\begin{vmatrix} -x & 1 & 0 & 0 \\ 1 & -x & 1 & 0 \\ 0 & 1 & -x & 1 \\ 0 & 0 & 1 & -x \end{vmatrix} = 0 \tag{2-58}$$

By this device we have, in a sense, reduced the secular determinant to being a function not, explicitly, of ε, anymore, but of this quantity, x. But, of course, if we find values of x which satisfy (2-58), we find values of ε which satisfy (2-55), since if x_I is the Ith root of (2-58) then the energy of the corresponding MO would be, according to (2-57),

$$\varepsilon_I = \alpha + x_I \beta \tag{2-59}$$

We have derived equation (2-58) in a detailed, step-by-step fashion in the present discussion, in order to make clear the arguments and assumptions underlying its formulation. In practice, however, if one wanted to do a calculation of this sort on an arbitrary hydrocarbon one would actually *start* by writing down equation (2-58)—"on the back of a postage stamp", so-to-speak—*merely on the basis of the σ-bond connectivity of the conjugation network in question.* For, in every case, on the present assumptions, $-x$ is going to appear everywhere on the diagonal of the secular determinant. Furthermore, once the atoms of the conjugated system have been arbitrarily[N12] numbered, all off-diagonal elements, $(i - j)$, of the secular determinant (2-58) are zero unless atom i is adjacent to (*i.e.*, in the σ-framework, is "bonded" to) atom j, in which case the $(i - j)$-element is unity[N13].

Expanding the determinant in equation (2-58) along the first row, we obtain

$$-x \begin{vmatrix} -x & 1 & 0 \\ 1 & -x & 1 \\ 0 & 1 & -x \end{vmatrix} + (-1)\cdot 1 \begin{vmatrix} 1 & 1 & 0 \\ 0 & -x & 1 \\ 0 & 1 & -x \end{vmatrix} = 0 \tag{2-60}$$

i.e.,

$$-x\{(-x)[(-x)(-x) - (1)\cdot(1)] + 1\cdot(-1)[1\cdot(-x) - (0)\cdot(1)]\}$$
$$+ (-1)\{1[(-x)(-x) - (1)\cdot(1)] + 1\cdot(-1)[(0)\cdot(-x) - (0)\cdot(1)\} = 0 \tag{2-60a}$$

which reduces to

$$x^4 - 3x^2 + 1 = 0, \tag{2-61}$$

from which we immediately deduce that

$$x^2 = \frac{+3 \pm \sqrt{5}}{2} \tag{2-62}$$

and (to three decimal places) the four roots $\{x_I\}$, $I = 1$–4, are $\pm 1{\cdot}618$ and $\pm 0{\cdot}618$; finally, from (2-59) we see that the energies, $\{\varepsilon_I\}$, $I = 1$ to 4, of the four molecular-orbitals will be[N14]

$$\alpha \pm 1{\cdot}618\beta$$

$$\alpha \pm 0{\cdot}618\beta$$

As anticipated, we have of course ended up with four MO's (and thus four ε_I's) having started with four basis atomic-orbitals in the LCAO-MO scheme. It would be useful to be able to arrange them into some kind of sequence and so we ought to give some thought to the *signs* of the quantities α and β. Since α effectively measures the energy of a π-electron in a carbon atom, this will be negative, for energy has to be expended in order to remove this electron. So α is certainly negative otherwise we could not talk about the attraction of the atom for an electron. The same argument suggests that since the bond attracts an electron β will also be negative. It is true that we do not know anything about the absolute *magnitudes* of α and β—this will be discussed in the next chapter. What we are trying to do here, by establishing the *signs* of α and β, is to set up a schematic pattern for the four MO-energies. The four energy-levels are conventionally represented diagrammatically as in Fig. 2-7.

We have what may be regarded as a natural zero, α, from which other energies may be measured, either "up" or "down", depending on whether they are above or below the line representing the energy, α; we shall thus have four horizontal lines, arranged vertically, schematically representing the four MO-energies, $\{\varepsilon_I\}$, $I = 1$–4. Since β is negative, the *lowest* energy (let us call it the energy of MO Ψ_1) will be the one with *largest* coefficient of β—*i.e.*, the largest x_I (in this case 1.618) amongst $\{x_I\}$, $I = 1$–4. The next-lowest, Ψ_2, is in this example $\alpha + 0{\cdot}618\beta$; then come $\alpha - 0{\cdot}618\beta$ and $\alpha - 1{\cdot}618\beta$ which we shall call the energies of MO's Ψ_3 and Ψ_4, respectively (Fig. 2-7). In the expression (equation (2-59))

$$\varepsilon_I = \alpha + x_I\beta$$

for the energy of the Ith orbital, if x_I (the coefficient of β) is positive (as is the case with Ψ_1 and Ψ_2, above) the orbital in question is referred to as a *bonding orbital* because the presence of an electron in such an orbital decreases the

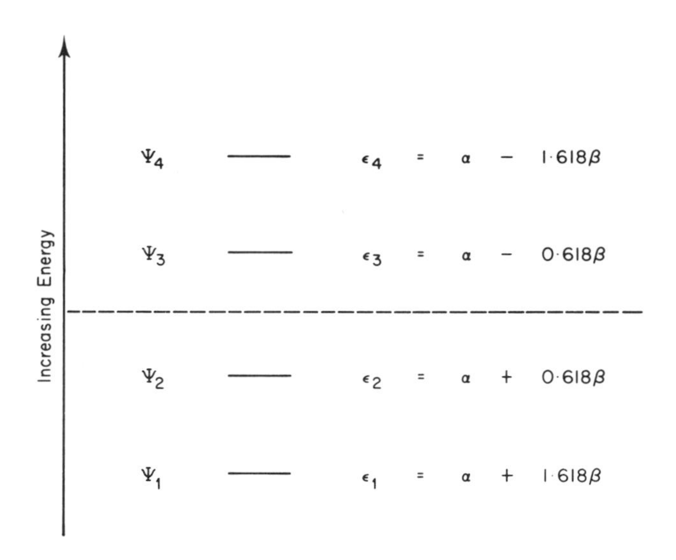

FIG. 2-7 Energy levels of the molecular orbitals of butadiene.

total π-electron energy of the system; those orbitals for which x_I is negative are called *anti-bonding* since, if one of *these* π-orbitals is occupied by an electron, the total π-electron-energy of the system is increased. If $x_I = 0$ is a root of the secular determinant then the molecular orbital (of energy $\varepsilon_I = \alpha$) associated with this root is referred to as a *non-bonding orbital.* There are no non-bonding orbitals in butadiene, but when they do arise they are very important in Hückel theory for they are often associated with triplet states (two electrons with parallel (*i.e.*, unpaired) spins) and thus with instability in molecules[R2]. We shall encounter them presently (in Chapter Five) when discussing the Hückel "$4p + 2$" Rule for monocyclic, conjugated systems. Having obtained the energy levels, $\{x_I\}$, $I = 1$–4, we now need to find the set of LCAO combinatorial coefficients, $\{c_{Ir}\}$, $r = 1$–4, corresponding to each suffix, I ($I = 1$–4, in turn). As an example, let us find the set, $\{c_{1r}\}$, $r = 1$–4, corresponding to $x_1 (= 1{\cdot}618)$. Substituting $x = 1{\cdot}618$ back into equation (2-58) and re-inserting the implied coefficients $\{c_{1r}\}$, we obtain

$$\left.\begin{aligned}(-1{\cdot}618)c_{11} + c_{12} + 0\cdot c_{13} + 0\cdot c_{14} &= 0\\ c_{11} + (-1{\cdot}618)c_{12} + c_{13} + 0\cdot c_{14} &= 0\\ 0\cdot c_{11} + c_{12} + (-1{\cdot}618)c_{13} + c_{14} &= 0\\ 0\cdot c_{11} + 0\cdot c_{12} + c_{13} + (-1{\cdot}618)c_{14} &= 0\end{aligned}\right\} \qquad (2\text{-}63)$$

Elimination from equations (2-63) shows the ratio $c_{11}:c_{12}:c_{13}:c_{14}$ to be 1:1·618:1·618:1. To obtain the absolute values of $\{c_{1r}\}$, $r = 1$–4 from these ratios we invoke the normalisation condition (2-34) to find that

$$\begin{aligned} c_{11} &= 0{\cdot}37 = c_{14} \\ c_{12} &= 0{\cdot}60 = c_{13} \end{aligned} \qquad (2.64)$$

Now take the root $x_2(=0{\cdot}618)$, and find the LCAO-coefficients $\{c_{2r}\}$, $r = 1$–4, corresponding to this by forming, from (2-58), another set of equations analogous to (2-63)—*i.e.*,

$$\left.\begin{aligned} (-0{\cdot}618)c_{21} + c_{22} + 0\cdot c_{23} + 0\cdot c_{24} &= 0 \\ c_{21} + (-0{\cdot}618)c_{22} + c_{23} + 0\cdot c_{24} &= 0 \\ 0\cdot c_{21} + c_{22} + (-0{\cdot}618)c_{23} + c_{24} &= 0 \\ 0\cdot c_{21} + 0\cdot c_{22} + c_{23} + (-0{\cdot}618)c_{24} &= 0 \end{aligned}\right\} \qquad (2\text{-}65)$$

Elimination from these equations and application of the normalisation condition leads to

$$\begin{aligned} c_{21} &= 0{\cdot}60 = -c_{24} \\ c_{22} &= 0{\cdot}37 = -c_{23} \end{aligned} \qquad (2\text{-}66)$$

Similarly, by substituting $x_3(=-0{\cdot}618)$ and $x_4(=-1{\cdot}618)$ in turn into the secular determinant, $\{c_{3r}\}$, $r = 1$–4, and $\{c_{4r}\}$, $r = 1$–4, may be found. The four LCAO-MO's of butadiene, $\{\Psi_I\}$, $I = 1$–4, thus turn out to be[N15]

$$\begin{aligned} \Psi_4 &= 0{\cdot}37\phi_1 - 0{\cdot}60\phi_2 + 0{\cdot}60\phi_3 - 0{\cdot}37\phi_4 \quad &&(\text{with energy } \alpha - 1{\cdot}618\beta) \\ \Psi_3 &= 0{\cdot}60\phi_1 - 0{\cdot}37\phi_2 - 0{\cdot}37\phi_3 + 0{\cdot}60\phi_4 \quad &&(\text{with energy } \alpha - 0{\cdot}618\beta) \\ \Psi_2 &= 0{\cdot}60\phi_1 + 0{\cdot}37\phi_2 - 0{\cdot}37\phi_3 - 0{\cdot}60\phi_4 \quad &&(\text{with energy } \alpha + 0{\cdot}618\beta) \\ \Psi_1 &= 0{\cdot}37\phi_1 + 0{\cdot}60\phi_2 + 0{\cdot}60\phi_3 + 0{\cdot}37\phi_4 \quad &&(\text{with energy } \alpha + 1{\cdot}618\beta) \end{aligned} \qquad (2\text{-}67)$$

Hence, equation (2-67) contains the complete solution to our problem—the four LCAO-MO's of butadiene and their energies.

2.8 Assigning Electrons to Molecular Orbitals: the *Aufbau* Principle

We might at this point ask ourselves what information of chemical interest is available from the data contained in equations (2-67). In an attempt partially to answer this we start by recalling that each π-MO in butadiene has been constructed from four atomic orbitals in an LCAO-scheme. Each of these atomic orbitals contributed one electron to the π-MO extending over the entire molecule and so these four electrons must be contained in, and by some criterion must be assigned to, the four MO energy-levels calculated for the molecule as a whole. The criteria adopted for this are the *Aufbau Principle* and the *Pauli Exclusion-Principle* used in a way exactly analogous to their application to atoms. The four π-electrons will therefore be assigned to the two orbitals of lowest energy, each of which will, according to the above Principles, contain two electrons of opposed (*i.e.*, "paired") spins. This situation is represented schematically in Fig. 2-8 which depicts

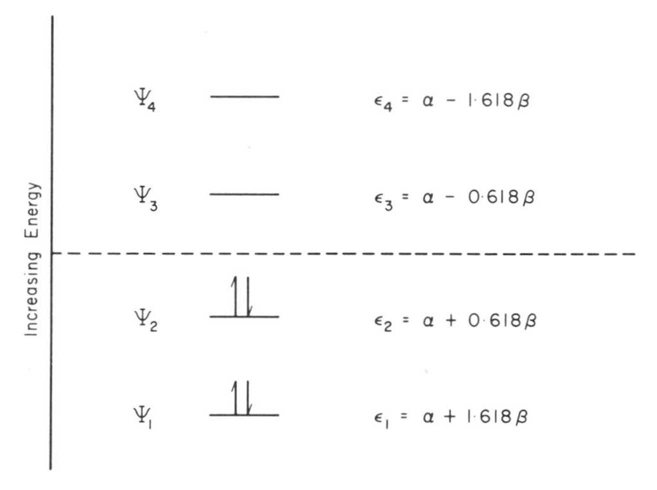

FIG. 2-8 The ground-state electronic-configuration of butadiene.

what, according to this model, would be regarded as the *ground state* for butadiene. Note that the two MO's of highest energy (Ψ_3 and Ψ_4) are *unoccupied* in the ground state. The ground-state electronic configuration of butadiene could therefore be written, in a shorthand notation, $(\Psi_1)^2(\Psi_2)^2$.

2.9 Total π-Electron Energy and Delocalisation Energy

Having established the ground-state electronic-configuration we may now say something about the total π-electron-energies of butadiene. Within the framework of Hückel theory, in which we have averaged the electron interactions with one another in the form of an effective Hamiltonian, we may regard the total π-electron-energy as simply the sum of the energies of the electrons in the occupied orbitals. Thus, in butadiene, we see from Fig. 2-8 that the total π-electron-energy, E^π, is just

$$E^\pi = 2(\alpha + 1{\cdot}618\beta) + 2(\alpha + 0{\cdot}618\beta) = 4\alpha + 4{\cdot}47\beta \qquad (2\text{-}68)$$

It would be of interest to relate this to what we conventionally describe as the resonance energy, or the delocalisation energy, of the molecule.

In order that we may do that we must consider for a moment an *isolated* double-bond and imagine doing exactly the same kind of calculation for this as we have just been doing for butadiene. Of course, this is a great deal easier than the butadiene calculation for we have to deal with only two atomic-orbitals; call them "1" and "2" (Fig. 2-9). Since there are only two atomic-

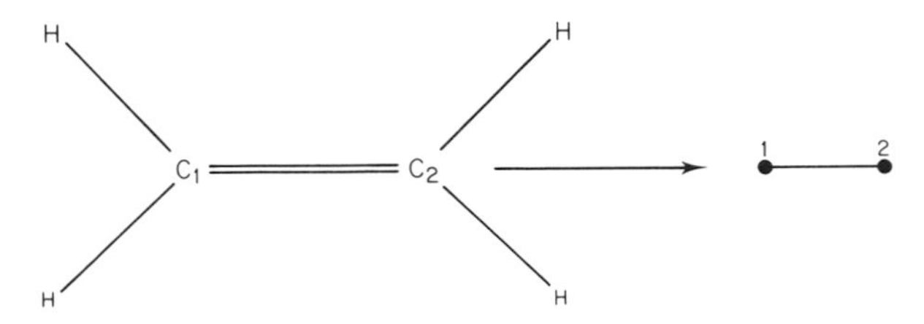

FIG. 2-9 Structural formula of ethylene and its equivalent molecular-graph representing the σ-bond connectivity of its carbon atoms.

orbitals, it is quite straightforward to show that the normalised MO of higher energy, Ψ_2, is $(1/\sqrt{2})(\phi_1 - \phi_2)$, with energy $(\alpha - \beta)$, and the one of lower energy, Ψ_1, is $(1/\sqrt{2})(\phi_1 + \phi_2)$, with energy $(\alpha + \beta)$. The two electrons in the π-electron system (one from each basis-AO) will be assigned, with opposite spins, to the lower orbital, Ψ_1, and so we have, schematically Fig. 2-10. We should therefore say that the π-electron energy of a single, isolated double-bond, on this model, is $2(\alpha + \beta)$. So it is relative to the energy of a *group* of such isolated double-bonds that we shall discuss the π-electron energy of butadiene; we shall call such a relative energy the *delocalisation energy* of the π-electron system. If we consider the traditional picture of

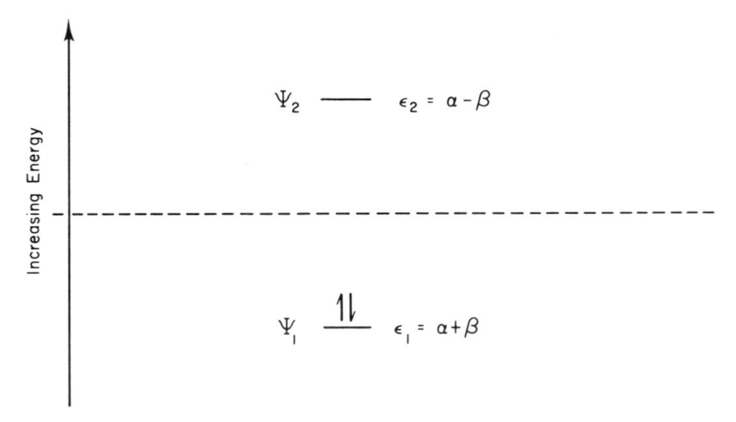

FIG. 2-10 Ground-state electronic-configuration of ethylene.

butadiene with fixed single- and double-bonds we would have what might be called, with an extension of the original usage, a Kekulé structure which involves an arrangement of two localised double-bonds—*i.e.*, Fig. 2-11.

1 2 3 4

FIG. 2-11 A Kekulé structure of butadiene, showing localised single- and double-bonds.

Since we have just shown that, on the present model, the π-electron energy of an isolated double-bond is $2\alpha + 2\beta$, the total π-energy of a classical Kekulé-structure such as that shown in Fig. 2-11, which contains two of these isolated double-bonds, is $4\alpha + 4\beta$. But we have shown previously (equation (2-68)) that the actual π-electron-energy of butadiene on the Hückel model is $4\alpha + 4{\cdot}47\beta$. We therefore interpret the difference between these two quantities ($0{\cdot}47\beta$ in this case) as being the result of allowing the electrons to delocalise, that is to say, allowing them to move in an MO over the complete framework of the molecule rather than being, as-it-were, "tied up"—two at one end of the molecule and two at the other—in localised bonding as in the Kekulé form of Fig. 2-11. This additional energy is sometimes called the *resonance energy* of the molecule but we shall prefer to call it the *delocalisation energy* of the π-electron system in question.

This kind of calculation can in principle be carried through for any type of conjugated system. When particularly large delocalisation-energies arise we might say that this is a characteristic of a particularly stable sort of system, whereas unstable or non-existent systems might be expected to have very small or even negative delocalisation-energies. We may take as an example of the first class, benzene, the prototype "aromatic" molecule. Since there are six basis-AO's, there are six LCAO-MO's and there are thus six roots to the 6×6 secular-determinant of the system. These turn out to be $(\alpha \pm 2\beta)$ and $(\alpha \pm \beta)$, the latter-two roots occurring *twice*[N16]. The six π-electrons are fed in, *Aufbau*-like, into the three lowest MO's, as in Fig. 2-12.

All the orbitals below the "α"-level, (*i.e.*, all the bonding orbitals) are seen to be completely filled. The fact that one pair of them is degenerate will not matter at this stage, although for other purposes it may be very important indeed. Thus,

$$E^{\pi}(\text{benzene}) = 2(\alpha + 2\beta) + 4(\alpha + \beta)$$
$$= 6\alpha + 8\beta \qquad (2\text{-}69)$$

The energy of the three isolated double-bonds in a benzene Kekulé-structure is

$$E^{\pi}(\text{Kekulé}) = 6\alpha + 6\beta \qquad (2\text{-}70)$$

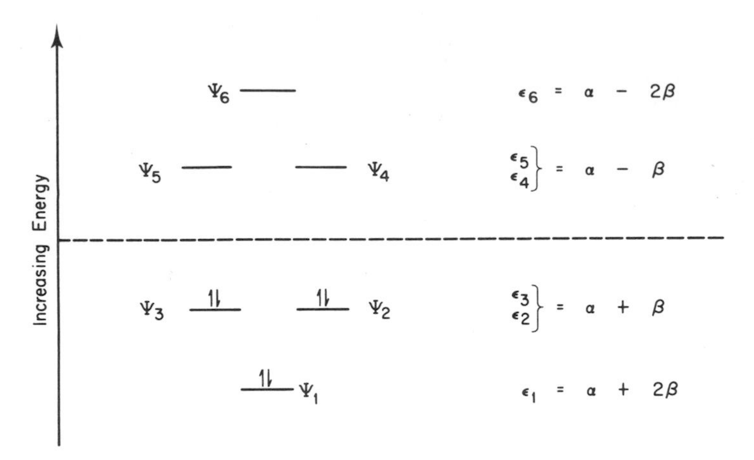

FIG. 2-12 Ground-state electronic-configuration of benzene.

Thus, the delocalisation energy for benzene is 2β. Notice that this is much larger than the $0{\cdot}47\beta$ obtained previously for butadiene. In fact we shall see later (Chapter Five) that ring systems very frequently have a much larger delocalisation-energy than the corresponding chain-molecule. This may be taken as some kind of heuristic verification of the experimentally observed fact that "aromatic" systems are stable and undergo substitution rather than addition-reactions, etc., while the corresponding chain-molecules will do both, depending on experimental conditions. This simple calculation thus suggests one reason why molecules whose conjugated systems comprise rings should be so different from open-chain molecules. There are other reasons, of course, to which we shall return later, but this is one.

2.10 Interpretation of the Physical Significance of the Butadiene-MO's

Having established the delocalisation energy of butadiene and compared it with that of benzene, let us now return to the LCAO-MO's of butadiene in equation (2-67) and proceed to a deeper interpretation of their physical significance. Consider the one of lowest energy, Ψ_1; it is made up of atomic orbitals, ϕ_r. Each atomic orbital is of large magnitude near the atom with which it is associated and is relatively small anywhere else. Consequently, the orbital, Ψ_1, in the vicinity of atom 1, will be given approximately by the expression $0{\cdot}37\phi_1$; notice that the coefficient of ϕ_1 has a positive sign. Near atom 2, Ψ_1 is approximately $0{\cdot}60\phi_2$ and again there is a positive sign in the coefficient. In fact, in Ψ_1, the coefficient of each $\{\phi_r\}$, $r = 1$–4, is positive. We may say, therefore, that Ψ_1 does not change sign along the length of the molecule; more technically, we can say that, along the length of the chain, Ψ_1 *has no nodes.* Of course, it has a node in the sense that all orbitals of π-symmetry have a node with respect to the plane of the molecule—reflection in the molecular plane produces a change of sign and that necessarily implies a zero value *in* the plane itself. At the moment, however, we are not thinking about nodes of this sort but rather of nodes which may occur in the various Ψ_I's as we proceed *along the chain* of the carbon atoms in butadiene. The same situation will clearly not hold for the other $\{\Psi_I\text{'s}\}$, $I = 2\text{–}4$, as a glance at equation (2-67) will confirm. Consider, for example, Ψ_2; as we proceed from atom 2 to atom 3, Ψ_2 changes from its value near atom 2 of *ca.* $+0{\cdot}37\phi_2$ to *ca.* $-0{\cdot}37\phi_3$ near atom 3—*i.e.*, Ψ_2 has changed sign between these two atoms. We may represent this symbolically by inserting a dotted line between atoms 2 and 3, as in Fig. 2-13, and agreeing that this should represent a node.

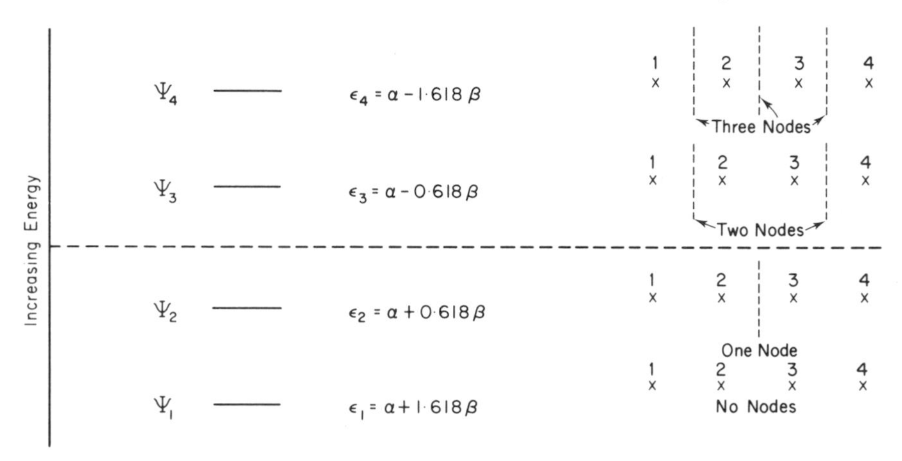

FIG. 2-13 π-Electron energy-levels of butadiene and the number of nodes in the corresponding molecular-orbitals.

If we were interested in the behaviour of Ψ_3 and Ψ_4 (of (2-67)), which would be involved in excited states of the molecule, we would observe that, in Ψ_3, there is a node between atoms 1 and 2, and between atoms 3 and 4, as shown in Fig. 2-13; and, in Ψ_4, there are nodes everywhere, as is evident from equation (2-67) and Fig. 2-13.

Now this discussion of nodes is not just a purely geometrical affair; it is going to tell us something about the bonding characteristics of these particular molecular-orbitals. Recall that, in the case of ethylene,

$$\frac{1}{\sqrt{2}}\phi_1 + \frac{1}{\sqrt{2}}\phi_2$$

was quite definitely a bonding orbital in the sense that electrons occupying it have an energy *lower* than α (*i.e.* the energy of this orbital involves a *positive* coefficient of β, which is itself negative) whilst the other ethylenic orbital,

$$\frac{1}{\sqrt{2}}\phi_1 - \frac{1}{\sqrt{2}}\phi_2,$$

was unambiguously anti-bonding. It looks, therefore, as if the presence of nodes can in some way be correlated with anti-bonding characteristics; when there are no nodes we would expect the circumstances to be favourable to bonding. (We are of course referring here to bonding *locally*—not with respect to the molecule as a whole but with respect to a particular region of the molecule.) On this basis we could then argue that Ψ_1, the lowest molecular-orbital in butadiene, will cause bonding between each pair of carbon atoms—*i.e.*, in all three regions, 1-2, 2-3, and 3-4 (see Fig. 2-13). Ψ_2, on the other hand, has one node and so it will be anti-bonding between atoms 2 and 3; it will, however, bond in regions 1-2 and 3-4. If we wish, we can extend this argument to Ψ_3 and Ψ_4 which are unoccupied (*i.e.*, contain no electrons) in the ground state. Ψ_4 is seen to be anti-bonding everywhere. Thus, an electron assigned to that orbital would try to weaken—indeed it would succeed in weakening—every one of the three bonds.

These arguments about nodes can also lead to some inferences about what would happen if electrons were excited from one orbital to another or even removed (ionised) from the system completely. Suppose, for example, that we caused an electron to be ionised from Ψ_2—as indeed is possible, experimentally; this will remove some bonding between 1 and 2, thus making the 1-2 bond weaker and, presumably, longer. On the other hand, this process removes some anti-bonding between 2 and 3 and the 2-3 bond will thus become stronger, and hence shorter, than before. Ionising an electron from Ψ_2 would therefore have the effect of strengthening the middle bond at the expense of the outer ones.

Now let us suppose that instead of ionising an electron we had excited an electron *from* Ψ_2 *to* Ψ_3. The electronic promotion, $\Psi_2 \rightarrow \Psi_3$, can be brought about experimentally be means of ultra-violet light in a standard photochemical process. Removal of an electron from Ψ_2 weakens the bond, 1-2—but then adding this electron to Ψ_3 *also* causes (further) antibonding in the 1-2 region; the end bonds (1-2 and 3-4) should thus be doubly weakened by this process. Between carbon atoms 2 and 3, anti-bonding is reduced and bonding is increased by this $\Psi_2 \rightarrow \Psi_3$ excitation, so that the central (2-3)-bond will be strengthened as a result of it. This could be illustrated diagrammatically, in a crude way, as in Fig. 2-14. Now Fig. 2-14 is something of an exaggeration but if the result of the ionisation process is depicted in this way it is clear that the end atoms, 1 and 4, will be preferentially reactive

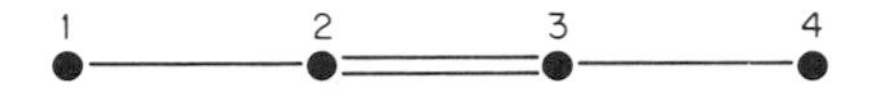

FIG. 2-14 Diagrammatic representation of the first excited-state of butadiene.

and polymerisation reactions, for example, should take place at these positions.

2.11 A Possible Improvement to the Original Hückel-Assumptions

Let us now go back to consider once more the ground state of butadiene, in which there are two electrons in Ψ_1 and two in Ψ_2, (Fig. 2-8) and use arguments similar to the ones we have just outlined in the context of ionisation and excitation processes to suggest certain improvements in the assumptions we have so far adopted in our development of the simple HMO-method. Looking again at the LCAO coefficients in equations (2-67) and the nodes in Fig. 2-13, we see that in the ground state the bond between carbon atoms 1 and 2 is strengthened by the presence of the two electrons in Ψ_1, and of the two electrons in Ψ_2. On the other hand, between carbon atoms 2 and 3 there is bond strengthening because of the presence of the two electrons in Ψ_1 but bond weakening due to the two electrons in Ψ_2. So, considerably weaker bonding is expected between C_2 and C_3 than between C_1 and C_2 (and, of course, by symmetry, the 3-4 bond is the same as the 1-2). Put in traditional nomenclature, therefore, the ground state of butadiene might be crudely described by the diagram in Fig. 2-11. Now if this ground-state picture is substantially correct—*i.e.*, if the end bonds really are more nearly double than is the middle bond—then it is natural to suppose that the *lengths* of these two types of bonds will be different. We shall be discussing the question of bond lengths in considerably more detail in a later chapter (Chapter Four). For the moment, however, we simply bear in mind that a double bond is shorter than a single bond and so we expect the end bonds (1-2 and 3-4) to be short and the middle bond (2-3) to be, by comparison, long. However, to conclude with this reasoning means that we have contradicted one of the assumptions with which we began, for when we started our Hückel-theory analysis we said that all resonance integrals between neighbouring carbon atoms were to have the same value, β. Now β, for the (1-2)-bond, for example, is (according to (2-49))

$$\beta_{12} = \int \phi_1 \mathscr{H} \phi_2 \, d\tau \qquad (2\text{-}71)$$

and hence, as the distance between the atoms 1 and 2 on which AO's ϕ_1 and ϕ_2 are centred *increases*, the absolute magnitude of β_{12} would be expected to *decrease*, numerically; (of course, β_{12} is actually negative but it is its absolute magnitude with which we are concerned for the moment). We ought to be able to say, therefore, that the magnitude of β for the central bond must be less than the magnitude of β for the end bonds—*i.e.*,

$$\|\beta_{23}\| < \|\beta_{12}\| \tag{2-72}$$

and, if we wanted to improve our calculation—to make it, in a sense, self-consistent with respect to the initial assumptions made about the resonance integrals—then we ought to take account of this. In order to do this, however, we require a knowledge of how β varies with bond length, and this is a topic dealt with in Chapter Four (§4.3c). For the moment we merely note that although this variation of resonance integral with bond length is of great importance for butadiene since it has two distinct types of carbon-carbon bonds within its conjugated system, the point is *not* relevant in the case of benzene because all six of its carbon-carbon bonds are equivalent. In benzene, therefore, all the bonds are of equal length and we can say with some justification and confidence that all its non-zero resonance integrals are equal; but in butadiene they are not and beginning the calculation with unequal β's for different bonds is something which we should bear in mind as a possible refinement to the HMO-treatment of butadiene and other systems in which not all carbon-carbon bonds are equivalent. We shall return to this point later in our discussion (Chapter Seven).

Three

The Hückel α and β Parameters and Heteroconjugated Systems

3.1 Introduction

It has been emphasised that one of the most-fortunate and notable features of Hückel theory is that many of its important conclusions are independent of the numerical values assigned to the parameters α and β. In the previous two chapters we have consequently developed the LCAO-scheme for the π-electron systems of conjugated molecules without actually specifying what numerical values must in practice be assigned to α and β; neither did we point out how the parameter α_r should be changed if the rth atom in the conjugated system is not a carbon atom. Although, as we shall see, the results of Hückel theory are, for most practical purposes, independent of the actual magnitude of α, (the carbon-atom Coulomb-integral for a carbon atom in benzene), not all the predictions to be obtained from the theory are independent of the value assumed for β. In the present chapter, we shall, therefore, examine (§3.2) some empirical ways of estimating β and compare the values thereby obtained. Then, in §3.3, we shall investigate how the Coulomb integral, α_r, of the rth atom should be modified if the carbon atom at this position in an arbitrary conjugated system is replaced by a hetero-atom (normally nitrogen or oxygen). We conclude the chapter by giving a worked example (§3.4) of how the secular determinant may be set up for a typical hetero-conjugated system (pyridine), and by commenting generally on the rôle played by α in Hückel calculations (§3.5).

3.2 Empirical values of β

(a) From Delocalisation Energies

In Chapter Two (§2.7) we saw that individual MO's have energies of the form

$$\varepsilon_I = \alpha + x_I \beta \qquad (3\text{-}1)$$

where x_I is one of the (at-most-n-distinct) roots of the secular determinant of the form shown in equation (2-58). The available π-electrons were then fed into the orbitals with these various energies and, by adding up the quantities, ε_I, for each of the π-electrons present, the total π-electron-energy for the system was obtained, as described in §2.9; when the corresponding quantity for a Kekulé structure was subtracted from this we obtained what we called the delocalisation energy for the molecule in question. Theoretical estimates of this quantity thus always emerge as some multiple of β, the standard carbon-carbon resonance-integral for a bond in benzene, as we saw in §2.9. Experimental estimates of the delocalisation energies of various conjugated hydrocarbons have been made from calorimetric measurements and hence, by comparing these for a series of molecules with the theoretical delocalisation-energies expressed in terms of β, an empirical value for this parameter may be obtained. Table 3-1 shows the theoretical and experimental delocalisation-energies for benzene, naphthalene, anthracene and phenanthrene (Fig. 3-1).

TABLE 3-1

	Delocalisation Energy		
Molecule	Theory	Experiment	Estimate of β
Benzene	$2{\cdot}00\beta$	37 kcal mol^{-1}	18·5 kcal mol^{-1}
Naphthalene	$3{\cdot}68\beta$	75 kcal mol^{-1}	20·4 kcal mol^{-1}
Anthracene	$5{\cdot}32\beta$	105 kcal mol^{-1}	19·7 kcal mol^{-1}
Phenanthrene	$5{\cdot}45\beta$	110 kcal mol^{-1}	20·2 kcal mol^{-1}

There are two conclusions which can be drawn from these numbers and which are typical of what can be obtained from other, similar systems. The first is that there is a very close parallel between the two columns of figures in Table 3-1; it appears, therefore, that these calculations (however crude they may be) are predicting at least a significant trend. Taking a value for β of the order of -20 kcal mol^{-1} leads to a reasonable agreement between these theoretical and experimental delocalisation-energies; thus, naphthalene is nearly twice as additionally stable, relative to a Kekulé structure, as is benzene, and anthracene is about three times so. The second point to observe from Table 3-1 is a more minor one and concerns the difference between anthracene and phenanthrene; anthracene may formally be considered as being derived from naphthalene by a *linear* annellation of a further benzene-ring, and phenanthrene by what we may refer to as a "zig-zag" annellation (Fig. 3-1). According to the theoretical predictions listed in Table 3-1, the "zig-zag" form is slightly more stable than the "straight" form—and this

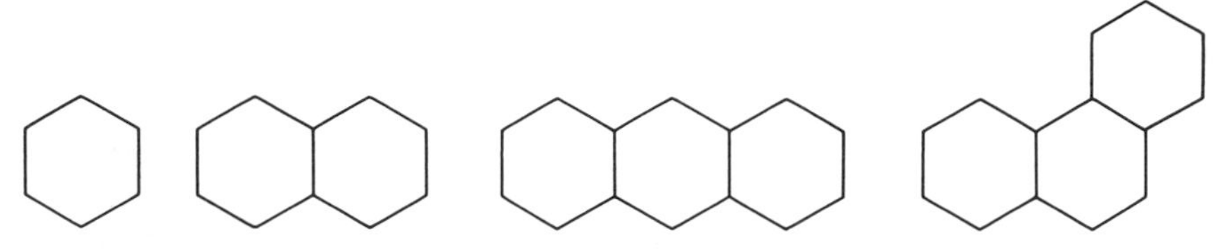

FIG. 3-1 Carbon-atom σ-bond connectivity of benzene, naphthalene, anthracene and phenanthrene.

is borne out by the experimental figures given in the table. Furthermore, this trend is not limited to these two molecules; for example, chrysene (Fig. 3-2a) is predicted (and found) to be more stable than tetracene (Fig. 3-2b). It appears, therefore, that we are beginning to obtain some interesting semi-quantitative results from a very simple method. Furthermore, experience has shown that the method very satisfactorily predicts the experimental delocalisation-energies of condensed, benzenoid, hydrocarbon-systems like these if a value of *ca.* -20 kcal mol^{-1} is assigned to β.

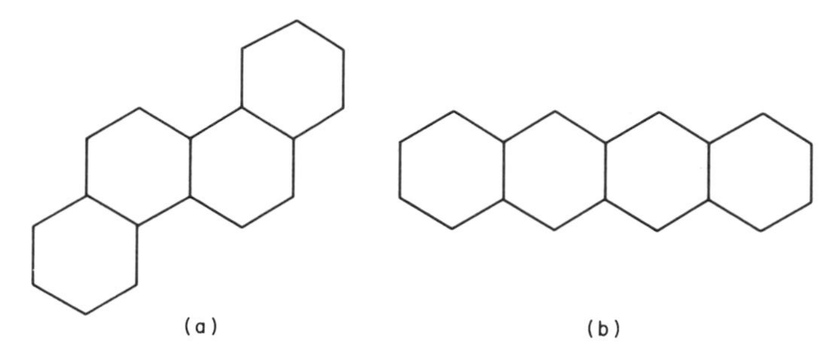

FIG. 3-2 Carbon-atom σ-bond connectivity of chrysene and tetracene.

(b) From Ionisation Potentials

Consider the ground state of a conjugated hydrocarbon—for example, benzene itself (Fig. 2-12). If we were interested in ionising an electron from this π-electron system, we could argue quite properly that it is an electron in the uppermost-occupied orbital which would be the easiest to remove in an ionisation process. Furthermore, the energy of the orbital from which the electron to be ionised is removed is some indication of the work which would be needed in order to effect such an ionisation. If the Ith orbital were the one in question we might then argue that the work involved in this process is given by equation (3-1). We could therefore very simply calculate the lowest ionisation potentials of a whole series of molecules—benzene, naphthalene, phenanthrene, butadiene, and so on—and Hückel theory would predict all of these ionisation potentials to be of the form $\alpha + x_I\beta$. Here, x_I is varying, in general, from molecule to molecule, whilst α and β are, of course, constant. Were we then to plot experimental ionisation-energy[R3] *vs* x_I for this series we would obtain a straight line of the standard form

$$y = mx + c \tag{3-2}$$

the slope, m, of which should represent the value of β. It turns out that the best least-squares straight-line through this set of points is

$$\text{Experimental Ionisation energy} = -(2{\cdot}48 \pm 0{\cdot}17)x_I - 7{\cdot}07\,(\text{eV}) \tag{3-3}$$

from which the approximate value of β is seen to be $-(2{\cdot}48 \pm 0.17)$ eV. Since the standard deviation in the slope is less than 15% this is quite a satisfactory correlation.

There is, however, something rather unsatisfactory about the empirical values of β we have obtained so far, something which is related to one of the basic defects of the Hückel method; it is that the value of β obtained from a comparison of theoretical and experimental ionisation-energies is more than twice that required to rationalise observed delocalisation-energies. This is the first time that we have so plainly met what will turn out to be a recurrent theme throughout Hückel theory—that the numerical parameters which are appropriate will vary with the physical phenomenon being considered. In the case of the ionisation potentials, the sequence of such potentials is very well determined; if one measured, for example, the ionisation potential for naphthalene and anthracene and estimated a value of β on that basis then the ionisation potential of the next linear-acene of the series, tetracene, would probably be very accurately determined. In fact, one would be fairly safe in identifying and predicting the ionisation potentials for a range of such molecules, once three or four had been obtained experimentally in order to calibrate the value of β for this property; but the value of β required to do this is of the order of twice that appropriate for delocalisation energies.

This situation may make some readers feel a little disappointed; however, let us examine it more closely for a little reassurance. As we saw in Chapter One, Hückel theory replaces the complete, π-electron Hamiltonian by a set of effective Hamiltonians for each electron; hence, whenever we have written $\mathscr{H}$ in Hückel theory it is really an effective Hamiltonian that we assign to electron 1, another to electron 2, and so on. We recall that this effective Hamiltonian for any given electron includes averaged Coulomb-repulsions from all the other electrons. Take, for example, $\mathscr{H}_{\text{effective}}$ for electron 1; it includes Coulomb repulsions from all the other electrons; in particular, let us focus attention on the repulsion between electron 2 and electron 1. Clearly we may say that the Coulomb interaction between electrons 1 and 2 appears in $\mathscr{H}(1)_{\text{effective}}$, albeit in a disguised, averaged way, and therefore it appears also in any energy obtained with this Hamiltonian. However, this *same* interaction, $1/r_{21}$, appears in $\mathscr{H}(2)_{\text{effective}}$, the effective Hamiltonian for electron 2—although it now appears, of course, as the force between electrons 1 and 2 instead of between electrons 2 and 1! We see therefore that when the energy contributed to the total by $\mathscr{H}(1)_{\text{effective}}$ is added to that contributed by $\mathscr{H}(2)_{\text{effective}}$ there is a sense in which we are counting the energy from the $(1/r_{12})$-term *twice*; a similar line of reasoning, of course, applies to the interaction between all other pairs of electrons in the π-system. This argument could be put more rigorously but the conclusion would still be that the value of β appropriate for the total energy is expected to be about half that suitable for any one orbital where, of course, electronic interactions of the type we have just discussed are *not* counted twice. We thus have at least a heuristic explanation for the factor of two, empirically observed between these two differently-determined values of β.

(c) From Proton-Magnetic-Resonance Measurements

Some of the most-recent estimates of the Hückel β-parameters have come from proton-NMR measurements of benzenoid hydrocarbons.[R4] When a conjugated hydrocarbon is in the presence of an external magnetic-field, the basis orbitals of its LCAO-MO's are no longer real; they can however be regarded as being derived from the $2p_{\pi}$-type atomic-orbitals on which Hückel theory is based, but multiplied by a complex exponential factor which depends on the external magnetic-field and the geometry of the molecule. In the presence of this external magnetic-field (in a direction perpendicular to the molecular plane and therefore parallel with the axis of the p_{π}-orbitals) the effect is classically almost as if the π-electrons of a conjugated system were constrained to move around the various rings of the conjugated network (in a preferential direction determined by the "up" or "down" nature of the magnetic-field vector). In so doing, these "mobile" π-electrons (constituting what are often—though somewhat evocatively—called "ring currents") create *secondary* magnetic-fields at the various peripheral-protons of a conjugated molecule. These secondary magnetic-fields in turn manifest themselves experimentally as shifts in the magnetic-resonance signals of these external hydrogen-atoms which can be detected and measured (in parts-per-million (τ) of the applied field) in an NMR-experiment. The details need not concern us; all we need to note for the purposes of the present discussion is that the secondary magnetic-field, B'_r, caused at the position of the rth peripheral hydrogen atom by the "ring currents" in all the various rings of a given molecule, is proportional to, and expressed explicitly in terms of, β. It is conventional to start by calculating the ratio B'_r/B'_{Benzene} which is thus independent of β (B'_{Benzene} is the similarly calculated secondary field at a benzene proton, due to the "ring current" in benzene); a regression is then performed between these ratios and a set of corresponding, experimental chemical-shifts, $\{\tau_r\}$, for certain of the various peripheral-protons, r, in a series of planar, polycyclic, condensed, benzenoid hydrocarbons. Such a correlation involving 66 non-equivalent protons in 16 different condensed, benzenoid molecules gave rise to

$$\tau_r = 1{\cdot}56\left(\frac{B'_r}{B'_{\text{Benzene}}}\right) + 4{\cdot}26\ (\text{ppm}) \qquad (3\text{-}4)$$

with a correlation coefficient of 96%. The slope of this line (1·56 ppm), together with the expression for B'_{Benzene} which explicitly involves β, yields a value for β of -35 kcals mol^{-1}. This value is different again from the other two discussed in this section because 1) a different phenomenon is being considered and 2) approximations involved in certain integrals peculiar to the magnetic theory, as well as other approximations introduced by the particular model chosen for the way in which the "ring currents" (calculated from simple-HMO wave-functions) affect the secondary field, B'_r, are *also* accommodated and compensated for by the regression—that is to say, by the value chosen for β. For example, if a different (and equally plausible) model is used to represent the way in which the same (HMO-calculated) "ring-currents" affect the secondary fields, the same experimental data give rise to

$$\tau_r = 1{\cdot}21\left(\frac{B'_r}{B'_{\text{Benzene}}}\right) + 4{\cdot}04 \qquad (3\text{-}5)$$

(correlation coefficient 98%) and a value of *ca.* $-42{\cdot}8$ kcal mol^{-1} for β.

The message is therefore clear: we do best to regard the simple Hückel theory as providing explanations of only a semi-quantitative nature. There is therefore no point in pressing it too far on the question of absolute magnitudes for β.

(d) Final Comment on the Selection of Values for β

One final word of caution may be sounded on the delocalisation energies discussed in §3.2a and tabulated in Table 3-1. They have been calculated without reference whatever to any changes in the σ-framework which may be required to go from a hypothetical Kekulé-structure to an actual, "delocalised", conjugated system—changes which, energetically, may sometimes be quite large. For example, to convert a benzene Kekulé-structure, with alternating bond-lengths of 1.34 and 1.51 Å, into a regular hexagon of side 1·40 Å would require the addition of *ca.* 30 kcal mol^{-1} to the basic σ-bond structure.[N17] We could therefore argue that because of this "compression energy", the true π-electron delocalisation-energy of benzene is not 37, but more like 76 kcal mol^{-1}. It will not have escaped the reader's notice that this would give rise to a value of β very nearly the same as that required for the interpretation of UV spectra . . . ! In view of the discussion in parts (a)–(d) of the present subsection, however, and particularly because of the arguments at the end of §3.2b, it is doubtful whether such refinements really are justified.

3.3 The Introduction of Hetero-atoms into a Conjugated System

(a) Effect on the Coulomb Integrals

It will be recalled from equation (2-16) that the Coulomb integral for an atom r is

$$\alpha_r = \int \phi_r \mathscr{H} \phi_r \, d\tau \qquad (3\text{-}6)$$

Since it may be regarded as a measure of the electron-attracting ability of atom r, α_r is evidently related to the electronegativity of this atom. It would be natural, therefore, to compare two atoms in a conjugated system—for example, nitrogen and carbon—by relating the difference in their Coulomb integrals to the difference in their electronegativities[R5] *i.e.*,

$$\alpha_r - \alpha_s = \text{constant} \times (\text{Electronegativity of Atom } r - \text{Electronegativity of Atom } s) \qquad (3\text{-}7)$$

Since the α-terms have the dimensions of energy (as also, of course, does the difference of electronegativities between atoms r and s) it is conventional to represent this difference as some (dimensionless) multiple, h, of β (referred to in the present discussion only as β_{CC}), to take α_s as the standard carbon-value (now called α_C) and hence to write

$$\alpha_r - \alpha_s = \alpha_r - \alpha_C = h\beta_{CC}, \quad \text{or} \quad \alpha_r = \alpha_C + h\beta_{CC} \qquad (3\text{-}8)$$

Thus, in equation (3-8), α_r is the Coulomb integral for some hetero-atom at the position labelled r in the conjugated system, α_C (what we have previously called just "α") is the standard Coulomb-integral of a carbon atom in benzene, and $\beta_{CC}(=\beta)$ is the standard resonance-integral of a carbon-carbon bond in benzene. β_{CC} is thus once more functioning here simply as a *unit* in terms of which energies are conveniently measured, and h should be proportional to the difference between the electronegativities of some atom, r, and of a carbon atom in benzene.

This represents a promising start to our introduction of hetero-atoms into a given conjugated hydrocarbon-system. We may further improve the scheme by observing that the value of h which is appropriate for a given atom will depend to a certain extent upon the rôle which that atom (r) is playing in the molecule. For example, if the rth atom were a nitrogen atom, we would not expect to find the same value of h for this atom in pyrrole as for the nitrogen atom in pyridine. The nitrogen atom in the pyridine-ring system may be thought of as being approximately sp^2-hybridised, as are the carbon atoms. Two of these nitrogen hybrids are used for bonding to the neighbouring carbon atoms and the third is occupied by two nitrogen σ-electrons to form a "lone pair" (Fig. 3-3a). Four of the valence electrons of

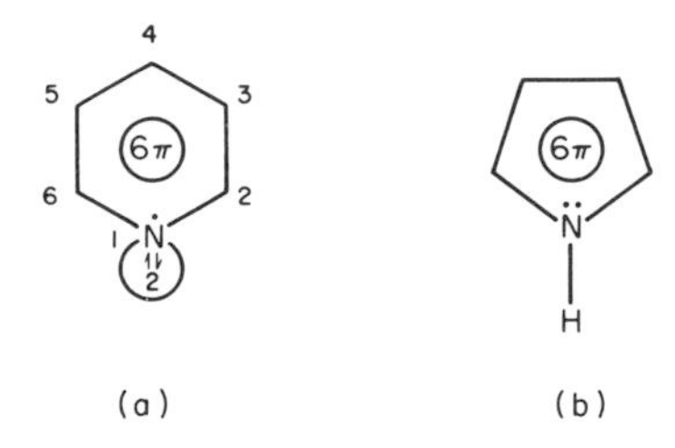

FIG. 3-3 π-Electron structure in (a) pyridine and (b) pyrrole.

nitrogen are thus involved in σ-orbitals and the final one is in an orbital which has π-symmetry with respect to the molecular framework. This p_π-orbital of nitrogen may effectively overlap with the p_π-orbitals of carbon and so the nitrogen atom in pyridine, like the carbon atoms, effectively contributes *one* electron to the π-system. In pyrrole (Fig. 3-3b), we may again imagine the nitrogen atom to be hybridised in an approximately trigonal fashion; this time three nitrogen-electrons are donated to these three nitrogen-hybrids in order to form bonds with two neighbouring carbon-atoms *and* (and here is where the nitrogen bonding in this molecule differs from that in pyridine) with a peripheral hydrogen-atom. Consequently, there are now *two* nitrogen valence-electrons left which may occupy an orbital of π-character; the nitrogen atom in pyridine thus donates *two* electrons to the π-system.

What we in fact require, therefore, is not the electronegativity as such, in the form in which Pauling[R5] referred to it; rather what we want, for use in equation (3-8), is something like an *effective* electronegativity which depends upon the way in which the atom in question fits into the molecular framework. For this reason, we are more interested in what could be called the *valence-state ionisation-potential*—that is to say, we imagine the atom, "prepared", as-it-were, in the way in which it will be needed in order to form the appropriate number of σ- and π-bonds (particularly the σ-bonds) and we then enquire about the ionisation potential of a π-electron in an atom of that kind. Clearly, this will be different in the case of (say) a nitrogen atom in pyrrole from its corresponding value in the case of a nitrogen atom in pyridine. If one were able, in the former case, to remove all the nitrogen atom's π-electrons this would leave the resulting species with an enormous attraction for electrons, since *two* electrons would have been taken away from it; this therefore becomes a very electronegative nitrogen-atom. Consequently, α_N for the pyridine nitrogen will not be as different from α_C as α_N for the pyrrolic nitrogen will be.

Having established this point, let us now examine some values of h which have often been used in a variety of circumstances. It is a rather distressing situation that nearly everyone who uses this approach has his own "private" set of h-values! This is sometimes very tiresome for it means that calculations taken from different literature-sources have to be examined very closely before they are directly compared. Plausible values—that is all we are going to say in the way of commending them to the reader—are given in Table 3-2.

TABLE 3-2

Atom a	h_a
$\dot{N}$ (pyridinium nitrogen)	0·5
$\ddot{N}$ (pyrrole)	1·5
$\overset{+}{N}$	2·0
$\dot{O}$	1·0
$\ddot{O}$	2·0

A very thorough discussion of this parametrisation problem, together with what, on certain criteria, may be regarded as an "optimum" set of parameters to use in a variety of situations, are given in Streitwieser's book.[R6] It cannot be denied, however, that it is a weak point in the theory that these quantities are not uniquely and "cleanly" determined.

(b) Effect on the Resonance Integrals

So much for the Coulomb integrals; we may now inquire how the resonance integrals are affected when a carbon atom is replaced by a hetero-atom. The resonance integrals, it will be recalled, represent a measure of the attraction of the corresponding *bonds* for electrons. Various proposals have been made for determining these when hetero-atoms are present, but it has to be admitted that none of them has a very satisfactory theoretical basis. The most plausible one of all which uses certain hypervirial relationships[R7] failed miserably in certain cases; it is therefore probably best to regard the β-values we are about to present as entirely empirical. We are going to say simply that just as the $\{\alpha_r\}$ were stated, relative to some standard (the carbon-atom Coulomb-integral), as some multiple of β_{CC}, so β_{rs}, between two different atoms r and s, will be some multiple (which we shall now call k) of the standard carbon-carbon resonance-integral, β_{CC}. That is to say,

$$\beta_{rs} = k_{rs}\beta_{CC} \tag{3-9}$$

Thus, we might have k-values as shown in Table 3-3. Now, of course, one would not wish to have to defend these absolute values too strongly; however, if they are used consistently and applied to a series of similar molecules then,

TABLE 3-3

Bond r-s	$k_{r\text{-}s}$
$C—\dot{N}$	0·8
$C—\ddot{N}$	1·0
N—O	0·7
C—O	0·8
C=O	1·0

on the whole, reasonable results are obtained. As an illustration of this we might consider, for example, the purines and pyrimidines, a series of heterocyclic systems which are of great importance biochemically. Using the sets of parameters detailed in Tables 3-2 and 3-3 one can go through the calculations as described in the next section (§3.4) and obtain very sensible variations between, say, the purine- and pyrimidine systems. Hence, although there is a good deal of empiricism in the choice of the parameters, if one is consistent with them the calculations are of use. For example, they can tell us in what parts of these molecules charge (Chapter Four) will tend to be concentrated and this will enable the prediction of sites which are chemically reactive and those which are likely to be of high biological activity. We must not therefore entirely despise this kind of semi-empiricism.

3.4 The Secular Determinant of Pyridine: A Worked Example

As an illustration of the way in which hetero-atoms are, in practice, accommodated into the Hückel scheme, we set up here the secular determinant for pyridine. Let us (arbitrarily) base it on the atomic numbering of Fig. 3.3a, in which the nitrogen atom is labelled "1". Then, by use of the zero-overlap approximation (§2.3), and the assumption that matrix elements, H_{rs} $(r \neq s)$, between non-nearest neighbours are zero (§2.2), arguments similar to those outlined for butadiene in §2.7 lead to the following secular determinant for pyridine.

$$\begin{vmatrix} \alpha_N - \varepsilon & \beta_{CN} & 0 & 0 & 0 & \beta_{CN} \\ \beta_{CN} & \alpha_C - \varepsilon & \beta_{CC} & 0 & 0 & 0 \\ 0 & \beta_{CC} & \alpha_C - \varepsilon & \beta_{CC} & 0 & 0 \\ 0 & 0 & \beta_{CC} & \alpha_C - \varepsilon & \beta_{CC} & 0 \\ 0 & 0 & 0 & \beta_{CC} & \alpha_C - \varepsilon & \beta_{CC} \\ \beta_{CN} & 0 & 0 & 0 & \beta_{CC} & \alpha_C - \varepsilon \end{vmatrix} = 0 \qquad (3\text{-}10)$$

Using data from Tables 3-2 and 3-3, we set

$$\begin{aligned} \alpha_N &= \alpha + 0{\cdot}5\beta \\ \beta_{CN} &= 0{\cdot}8\beta \end{aligned} \qquad (3\text{-}11)$$

where α is now again written for α_C, and β simply stands, once more, for β_{CC}. The secular determinant for pyridine is then

$$\begin{vmatrix} 0{\cdot}5\beta + \alpha - \varepsilon & 0{\cdot}8\beta & 0 & 0 & 0 & 0{\cdot}8\beta \\ 0{\cdot}8\beta & \alpha - \varepsilon & \beta & 0 & 0 & 0 \\ 0 & \beta & \alpha - \varepsilon & \beta & 0 & 0 \\ 0 & 0 & \beta & \alpha - \varepsilon & \beta & 0 \\ 0 & 0 & 0 & \beta & \alpha - \varepsilon & \beta \\ 0{\cdot}8\beta & 0 & 0 & 0 & \beta & \alpha - \varepsilon \end{vmatrix} = 0 \qquad (3\text{-}12)$$

As in the case of hydrocarbons (§2.7), it is convenient to divide each row of the secular determinant by β (permissible since β is not zero, but the determinant is) and to express energies relative to α and in units of β—*i.e.* to set

$$x = \frac{\varepsilon - \alpha}{\beta} \tag{3-13}$$

These two processes turn (3-12) into (3-14)

$$\begin{vmatrix} 0{\cdot}5 - x & 0{\cdot}8 & 0 & 0 & 0 & 0{\cdot}8 \\ 0{\cdot}8 & -x & 1 & 0 & 0 & 0 \\ 0 & 1 & -x & 1 & 0 & 0 \\ 0 & 0 & 1 & -x & 1 & 0 \\ 0 & 0 & 0 & 1 & -x & 1 \\ 0{\cdot}8 & 0 & 0 & 0 & 1 & -x \end{vmatrix} = 0 \tag{3-14}$$

This determinant is thus seen to be precisely the secular determinant which would have been obtained for benzene, except that one diagonal element (the (1,1)-element) is $(0{\cdot}5 - x)$ instead of $-x$ and the (1,2)- and (2,1)- and the (1,6)- and (6,1)-off-diagonal elements are 0·8 instead of 1.

This determinant may now be expanded and solved in exactly the same way in which the corresponding determinant for butadiene was expanded and solved in §2.7. In the present case, development of the left-hand side of (3-14) will lead to a sixth-order polynomial in x, the roots of which will be $\{x_I\}$, $I = 1, 2, \ldots, 6$, which will be related (*via* (3-13)) to the six π-MO's of pyridine, with energies $\{\varepsilon_I\}$, $I = 1, 2, \ldots, 6 = \{\alpha + x_I\beta\}$, $I = 1, 2, \ldots, 6$. Each one of the six roots, x_I, when substituted back into (3-14), will give rise to the set of LCAO coefficients, $\{c_{Ir}\}$, $r = 1, 2, \ldots, 6$, for the corresponding MO, Ψ_I—or, rather, the *ratios* of these coefficients, the absolute magnitudes of which may finally be established by invoking the normalisation condition embodied in equation (2-34).

We see therefore that once we accept the empirical Coulomb-integrals and resonance-integrals alleged, by some criteria, to be appropriate for a given hetero-atom and for the bonds in which this hetero-atom is involved, it is a relatively simple matter to incorporate these into the secular determinant and thus to set up this determinant, for an arbitrary, hetero-conjugated system, on the basis of the Hückel approximations. Furthermore, once this secular determinant has been constructed, its solution to yield the energy levels and LCAO-MO's of the hetero-conjugated system follows as straightforwardly as in the case of the corresponding (homo-conjugated) hydrocarbon. In this discussion, and in that of §2.7, we have suggested what might be called "brute-force" methods for solving the secular determinant; in practice, however, the algebra may sometimes be simplified (as indeed it can in the pyridine case) by making appeal to the elements of symmetry extant in the conjugated system in question, and applying the methods of Group Theory; this, however, is the subject of Appendix B.

3.5 Postscript on the Value of α

In this chapter we have given some consideration to empirical estimates of β, the standard carbon-carbon resonance-integral for a carbon atom in benzene, and to the way in which Coulomb integrals are modified (in terms of β) if a carbon atom in a given conjugated system is replaced by a hetero-atom. The observant reader will no doubt have noticed that we have so far been careful to make no mention of the absolute magnitude of α, the Coulomb integral for a carbon atom in benzene, with reference to which all energies, and all Coulomb integrals for hetero-atoms, have so far been expressed. This is because, as was hinted at in the introduction to this chapter (§3.1), for most purposes we do not *need* to know it—at least in the neglect-of-overlap case; let us see why.

We have seen that, when overlap is neglected, the energy levels of a given conjugated system are the values of ε which satisfy an equation of the form:

$$|\mathbb{H} - \varepsilon\mathbb{1}| = 0 \tag{3-15}$$

We now recall that, on the basis of the approximations which we have used so far, if the conjugated system in question is a pure hydrocarbon then all the diagonal elements of $\mathbb{H}$ will be α; if a hetero-conjugated system is being dealt with then the diagonal elements of $\mathbb{H}$ will be of the form $\alpha + h_{a_r}\beta$ where the $\{a_r\}$, $r = 1, 2, \ldots, n$ represent various different atoms which may or may not be hetero-atoms; (if the rth atom were a carbon atom then, of course, on this model, h_{a_r} would be zero). Whatever the type of system, therefore, the term α, either alone (if the corresponding atom is a carbon atom) or with other terms (if the corresponding atom is a hetero-atom), will appear in *every* diagonal element of $\mathbb{H}$. Changing the value of α is thus equivalent to changing every diagonal element of $\mathbb{H}$ *equally*, whilst leaving the off-diagonal elements unaltered. It is clear from equation (3-15) that the only result of this process is to bring about an *exactly similar change in each of the values of* ε, $\{\varepsilon_I\}$, $I = 1, 2, \ldots, n$, which satisfy (3-15); in other words, we are thereby altering the zero-base with respect to which the eigenvalues of $\mathbb{H}$ are measured. It is clear from this argument that the eigen*vectors* of $\mathbb{H}$—*i.e.* the corresponding LCAO-MO coefficients—will *not* be affected by a change in the numerical value of α. Now, as we shall see in the next chapter, most of the properties of interest which can be calculated in Hückel theory (such as atomic changes, bond orders and free valences) are calculated from these *coefficients* and hence do not depend on the absolute value of α. Other properties accessible *via* Hückel theory such as transition energies, magnetic properties, and reactivities depend on the LCAO-coefficients and/or the *differences* between MO-energies and these also, therefore, are independent of the actual numerical value of the carbon-atom Coulomb-integral. The only property to which we shall make reference in this book which explicitly involves a numerical value for α is an ionisation potential of a molecule (discussed in §3.2b), since the expression for this (equation (3-1)) makes reference to the *absolute* energy of a molecular orbital. In fact, although we did not mention it in the discussion of §3.2b, the constant term (*ca.* $-7{\cdot}07$ kcal mol^{-1}) in the regression equation (3-3) might be interpreted as being an estimate of α.

In the next chapter we proceed to a discussion of atomic charges, bond orders and free valences, none of which depends on taking any explicit empirical value for α or β. In Chapters Five and Six we deal with the Hückel Rule of Aromaticity and the Coulson–Rushbrooke Theorem on Alternant Hydrocarbons, both of which are also independent of any numerical values assumed for these basic parameters.

Four

Charge, Bond Order and Free Valence

4.1 Introduction

We have so far seen how the Hückel LCAO-MO energy-levels and combinatorial coefficients may be obtained in the case of hydrocarbons (Chapter Two) and hetero-systems (Chapter Three) and how, by application of the *Aufbau* Principle, the π-electrons of the system may be assigned to the available orbitals in order to obtain the ground-state electronic configuration. In the present chapter, we shall therefore assume that we know (i) the ground state (or some other relevant electronic configuration) and (ii) the LCAO coefficients, of a given molecule and then proceed to define and calculate three specific indices of particular chemical interest—charge (§4.2), bond order (§4.3) and free valence (§4.4).

4.2 π-Electron Charge

(a) The Concept and Definition of π-Electron Charge

Let us consider the Ith MO

$$\Psi_I = c_{I1}\phi_1 + c_{I2}\phi_2 + \cdots + c_{In}\phi_n = \sum_{r=1}^{n} c_{Ir}\phi_r \qquad (4\text{-}1)$$

It would be physically reasonable to say that this wave function must determine, in some way, the probability of an electron in Ψ_I being associated with each particular atom, r, of the conjugated system—what we might rather intuitively call the charge density associated with that atom. Probabilities and charge densities are given in Quantum Mechanics by the *square* of the amplitude of a wave function; hence we should be interested, as far as charges are concerned, in $|\Psi_I|^2$. Now we know that, from the normalisation condition (equation (2-33)), in the neglect-of-overlap case,

$$\int|\Psi_I|^2 \, d\tau = \int(c_{I1}\phi_1 + c_{I2}\phi_2 + \cdots + c_{In}\phi_n)^2 \, d\tau$$

$$= c_{I1}^2 + c_{I2}^2 + \cdots + c_{In}^2 = \sum_{r=1}^{n} c_{Ir}^2 \qquad (4\text{-}2)$$

(since the neglect of overlap determines that the cross-terms must be zero). On squaring Ψ_I and integrating, therefore, we obtain a term c_{Ir}^2 associated with each atom r ($r = 1, 2, \ldots, n$) in the conjugated system. An electron in the orbital Ψ_I must of course be *somewhere*—and, indeed, the normalisation condition (equation (4-2)) does require that the total probability of its being somewhere ($\int |\Psi_I|^2 \, d\tau$) *is* unity. It is almost inevitable, therefore, that, under these circumstances, one should regard the individual terms c_{Ir}^2 as representing the probability that an electron in the MO Ψ_I shall be found on the particular atom r of the conjugated network. From here on we shall drop the term "probability" and refer to "charge density" and say that the electron is, as-it-were, "shared out" among the nuclei and that the "amount" of this electron which accrues to nucleus r from the orbital Ψ_I is given exactly by c_{Ir}^2.

Since we are saying that an electron in Ψ_I contributes an amount c_{Ir}^2 to the π-electron-charge on atom r, it follows that the *total π-charge* on this atom, denoted q_r, is the sum of such quantities from all π-electrons in all orbitals, *i.e.*,

$$q_r = \sum_{\substack{\text{over all} \\ \pi\text{-electrons}}} c_{Ir}^2 \qquad (4\text{-}3)$$

Another (and more conventional) way of writing this is to represent it as a sum over *orbitals*, rather than electrons, as in equation (4-4).

$$q_r = \sum_{I=1}^{n} \nu_I c_{Ir}^2 \qquad (4\text{-}4)$$

in which ν_I denotes the number of electrons assigned to the orbital Ψ_I. From this it follows that if the Ith orbital is doubly occupied, $\nu_I = 2$; if this orbital is singly occupied (a situation which we have encountered, for example, in radicals), $\nu_I = 1$, and if Ψ_I is unoccupied (empty) in the particular electronic configuration being considered, then $\nu_I = 0$.

(b) Uses of the Concept

Many important consequences follow from this simple concept and definition. Given an electronic configuration and a set of LCAO-coefficients

$\{c_{Ir}\}$, $I = 1, 2, \ldots, n$, $r = 1, 2, \ldots, n$, the set $\{q_r\}$, $r = 1, 2, \ldots, n$ may easily be calculated, from equation (4-4), for any arbitrary, conjugated system. Using this approach, we should begin to see quantitatively what, for example, the π-electron charge on the nitrogen atom in pyridine (§3.3, Fig. 3-3) would be. It would be more than one electron—but of course less than two—and this excess-charge on the nitrogen would be taken, in a preferential way, from the carbon atoms. We could therefore then obtain an indication of the charge distribution around the ring. The π-electron charge-densities of pyridine, calculated on the basis of the parameters discussed in §3.3 (Tables 3-2 and 3-3), are shown in Fig. 4-1. Such data will enable prediction of the

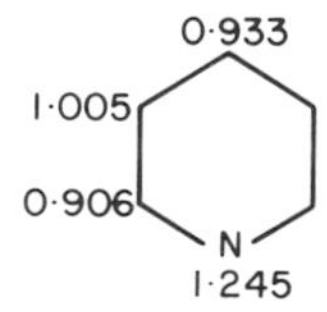

FIG. 4-1 π-electron charges in pyridine.

activity of pyridine in nucleophilic or electrophilic reactions. We can, however, go further than this; suppose that pyridine were ionised and that we knew from which orbital the ionised electron had been removed. We should then be able to say that atom r lost a given amount of charge in the process of ionisation; now, of course, different atoms will lose different amounts of charge (Fig. 4-2a). It is then possible to say which atoms lose the most charge—or, alternatively, if, instead of ionising an electron, one were to *add* a *further* electron to give an overall anionic species (Fig. 4-2b), it would

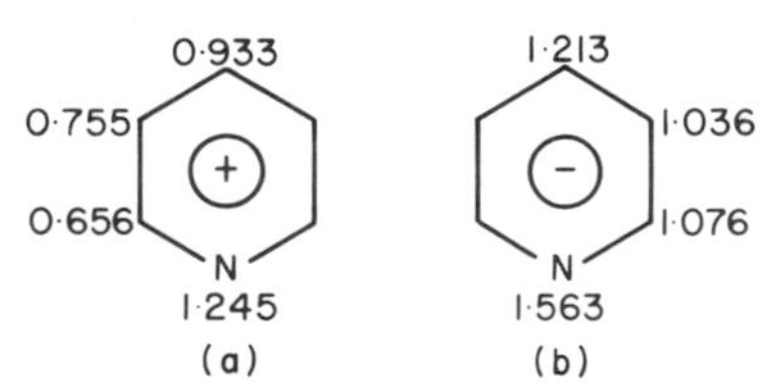

FIG. 4-2 Charge densities in the cationic (a), and anionic (b) forms of pyridine.

be evident which atoms gain the most charge. We thus begin to be able to discuss another entire range of chemical problems simply by referring to the changes in charge distribution which occur on addition or subtraction of an electron.

(c) Example: The Charge Densities of Butadiene

In order to illustrate practical application of equation (4-4) in a given case, let us consider calculating the π-electron charges on the four constituent carbon-atoms of butadiene (Fig. 2-6) in its ground-state, $(\Psi_1)^2(\Psi_2)^2$

TABLE 4-1

			r			
			1	2	3	4
Ψ_1	$\varepsilon_1 = \alpha + 1{\cdot}618\beta$	$2c_{1r}^2$	0·28	0·72	0·72	0·28
Ψ_2	$\varepsilon_2 = \alpha + 0{\cdot}618\beta$	$2c_{2r}^2$	0·72	0·28	0·28	0·72
	Total $\sum \nu_I c_{Ir}^2$ for $(\Psi_1)^2(\Psi_2)^2$:		1.00	1.00	1·00	1·00

(Fig. 2-8). We note from Fig. (2-8) that, in the ground state, $\nu_1 = \nu_2 = 2$ and $\nu_3 = \nu_4 = 0$; hence, all we require from equations (2-67) are $\{c_{1r}\}$, $r = 1, 2, \ldots, 4$ and $\{c_{2r}\}$, $r = 1, 2, \ldots, 4$. The application of equation (4-4) to these data is then summarised in Table 4-1. Thus we find that the π-electron charge-densities on all four carbon atoms of butadiene are unity. This is by no means fortuitous and is always the case for a certain class of molecules (called alternant hydrocarbons and dealt with in Chapter Six) to which butadiene belongs. The charge distributions in excited-state species will also be discussed in detail in the context of the Coulson–Rushbrooke Pairing-Theorem in §6.5.

4.3 Coulson Bond Order

(a) The Concept and Definition of Bond Order

In taking the Ith molecular orbital, Ψ_I (equation 4-1), squaring it and integrating, we obtained terms such as c_{Ir}^2 which were interpreted (§4.2) as representing the partial charge on the rth atom, due to an electron in Ψ_I. *Before* the integration is performed, however, there are in the expression for $|\Psi_I|^2$ *cross-terms* of the type $c_{Ir}c_{Is}\phi_r\phi_s$—as may be confirmed by expanding the integrand on the right-hand-side of equation (4-2). Let us consider the product $\phi_r\phi_s$; ϕ_r is the atomic orbital centred on atom r and it thus has a large value in the vicinity of atom r, and a small value, relatively, everywhere else; similarly, ϕ_s has a large value around atom s, but is insignificant in other regions of the molecule. The product of these two atomic orbitals can thus be of importance only in those parts of space which are near to both r and s—*i.e.*, in the region of the *bond* between atoms r and s. It is natural, then, to define a concept of *bond order*—or, more properly, *partial bond-order* since we are at present considering contributions from only one orbital, Ψ_I—on the basis of the cross-terms $c_{Ir}c_{Is}$. An electron in Ψ_I is thus considered to contribute $c_{Ir}c_{Is}$ to the π-bond-order between atoms r and s. This quantity will be particularly important if r and s are neighbours (*i.e.*, are already joined by a σ-bond) but will otherwise be of little value since the magnitude of ϕ_r and ϕ_s will never be large anyway. In a manner exactly analogous to that in which equation (4-3) was justified we can define the overall π-bond-order between atoms r and s (denoted p_{rs}) as the sum of these $c_{Ir}c_{Is}$-quantities over all electrons in all orbitals—*i.e.*

$$p_{rs} = \sum_{\substack{\text{over all} \\ \pi\text{-electrons}}} c_{Ir}c_{Is} \tag{4-5}$$

Again, it is more conventional to represent this as a sum over *orbitals*, rather than electrons, and to write the overall *π-bond-order*[N18], p_{rs}, as

$$p_{rs} = \sum_{I=1}^{n} v_I c_{Ir} c_{Is} \tag{4-6}$$

where v_I, the number of electrons occupying the Ith orbital in the particular electronic configuration under consideration, has the same significance as it had in equation (4-4).

There seemed to be a very plausible reason for introducing the concept of charge in the way in which we developed it in §4.2, but there is perhaps not such an immediately obvious reason for wishing to discuss this new concept, bond order. Let us therefore confirm that equation (4-6) is a sensible definition of such a quantity by considering the simplest-possible example, ethylene. It will be recalled from §2-9 that, in this molecule, there is only one occupied molecular-orbital, $\Psi_1 = (1/\sqrt{2})(\phi_1 + \phi_2)$, and it contains the two π-electrons of the system. We note that it is normalised, by virtue of the factor of $1/\sqrt{2}$ which premultiplies the bonding combination $(\phi_1 + \phi_2)$. For this particular orbital we thus have

$$c_{11} = c_{12} = \frac{1}{\sqrt{2}} \tag{4-7}$$

Furthermore, in the ground state, $(\Psi_1)^2(\Psi_2)^0$,

$$\left.\begin{aligned} v_1 &= 2 \\ v_2 &= 0 \end{aligned}\right\} \tag{4-8}$$

Application of equation (4-6) thus gives

$$p_{12} = \left(2 \times \frac{1}{\sqrt{2}} \times \frac{1}{\sqrt{2}}\right) + \left(0 \times \frac{1}{\sqrt{2}} \times \left(-\frac{1}{\sqrt{2}}\right)\right) = 1 \tag{4-9}$$

(Here, although coefficients of Ψ_2 are not really required since it is unoccupied, the (zero) contribution from the anti-bonding orbital $\Psi_2 = (1/\sqrt{2})(\phi_1 - \phi_2)$ has been, for the sake of clarity, explicitly included.) What we have just calculated is, of course, strictly only the π-electron contribution to the total bond-order of ethylene and it is therefore often written p_{12}^{π}. It follows quite naturally that, for the bond r-s, we have

$$p_{rs}^{\text{Total}} = p_{rs}^{\sigma} + p_{rs}^{\pi} \tag{4-10}$$

In the present example, there is one sigma-bond between carbon atoms 1 and 2 in ethylene and hence p_{12}^{σ} is 1; we have just seen that p_{12}^{π} for ethylene is also 1 and this therefore gives the familiar $p_{12}^{\text{Total}} = 2$, for the ethylenic double bond. The definition (4-6) is thus seen to be a rational one as far as what we might regard as the ordinary, isolated double-bond is concerned; it can also be shown to be satisfactory for the triple bond (as in acetylene).

(b) Example: Bond Orders for the Ground- and Excited States of Butadiene

Let us now examine a more realistic application of equation (4-6) to a particular molecule in which π-electron delocalisation may take place; we appeal again to butadiene since the LCAO-coefficients for it are already available in equation (2-67). The lowest MO, which we have always called Ψ_1, has the form

$$\Psi_1 = 0{\cdot}37\phi_1 + 0{\cdot}60\phi_2 + 0{\cdot}60\phi_3 + 0{\cdot}37\phi_4 \tag{4-11}$$

Let us first consider what contributions to the various π-bond orders (denoted $p_{rs}^{(1)}$) are made by *one* electron in this MO *alone*. $p_{12}^{(1)}$, for example, is given by

$$p_{12}^{(1)} = 0{\cdot}37 \times 0{\cdot}60 = 0{\cdot}22 \tag{4-12}$$

In the same way,

$$p_{23}^{(1)} = 0{\cdot}60 \times 0{\cdot}60 = 0{\cdot}36 \tag{4-13}$$

Equations (4-12) and (4-13) go considerably further than we were able to in §2-10, where we merely said that certain orbitals are bonding in some regions and anti-bonding in others. We now have actual numbers to assign to these bonding propensities and quantifying a previously qualitative concept is always an advantage, when it is possible. Hence we can now say that an electron in Ψ_1 is bonding in all three regions 1-2, 2-3, and 3-4, in the proportions 0·22:0·36:0·22, respectively. The same process can be carried out for Ψ_2 when the corresponding quantities 0·22, −0·13, 0·22 are obtained, as systematised in Table 4-2. It is important to observe the negative sign.

We have already noted in §2-10 that Ψ_2 has a node in the region between atoms 2 and 3 and now we are able to assign a magnitude to its bonding and anti-bonding contributions in various parts of the molecule. By adding

TABLE 4-2

Bond, rs		1-2	2-3	3-4
Ψ_1	$p_{rs}^{(1)}$	0·22	0·36	0·22
Ψ_2	$p_{rs}^{(2)}$	0·22	−0·13	0·22
$(\Psi_1)^2(\Psi_2)^2$	p_{rs}^{π}	0·89	0·47	0·89
Total Bond Order	p_{rs}^{Total}	1·894	1·447	1·894

the $p_{rs}^{(I)}$-terms in each column in Table 4-2, and remembering to multiply by 2 because Ψ_1 and Ψ_2 are both doubly occupied in the ground-state $((\Psi_1)^2(\Psi_2)^2)$, we obtain the overall π-bond-orders p_{rs}^{π} of the three bonds 1-2, 2-3, and 3-4, as shown in the Table, giving the values 0·89, 0·47 and 0·89, respectively. The *total* bond orders, p_{rs}^{Total}, within this Hückel definition (which, it must be admitted, is not wholly satisfactory but it will suffice for our present purposes) are thus 1·894, 1·447 and 1·894. We see, therefore, that the end bonds are more like double bonds (*i.e.*, like the ethylenic linkage

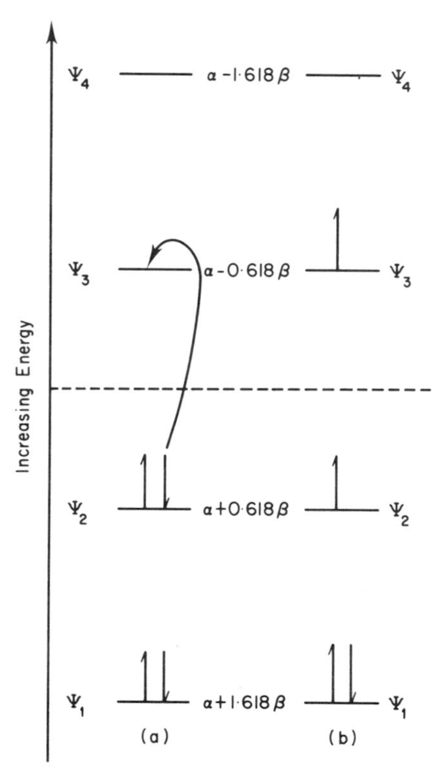

FIG. 4-3 Electron promotion to produce the first excited-state of butadiene.

with total bond-order of 2, as we have seen) than is the central bond so that the conventional picture for butadiene as Fig. 2-11 really does seem to be quite a rational one.

We might now investigate what would happen if we were to excite this molecule; let us suppose we excite an electron from the highest-occupied orbital in the ground state (Ψ_2) to the lowest-unoccupied orbital (Ψ_3)—*i.e.*, $\Psi_2 \rightarrow \Psi_3$ (Fig. 4-3a → 4.3b). This is the least-energetic and hence the longest-wave-length absorption in butadiene. By use of the LCAO-coefficients for Ψ_3 in equation (2-67), the partial bond-orders, $p_{rs}^{(3)}$, due to an electron in Ψ_3, are easily shown to be $-0{\cdot}22$, $+0{\cdot}13$ and $-0{\cdot}22$ for the bonds (r-s) 1-2, 2-3 and 3-4, respectively. Applying equation (4-6) to the configuration $(\Psi_1)^2(\Psi_2)^1(\Psi_3)^1$ using the coefficients of equations (2-67) then leads to

$$0{\cdot}45 \qquad 0{\cdot}72 \qquad 0{\cdot}45$$

for the overall π-bond-orders of this excited species, and thus

$$1{\cdot}45 \qquad 1{\cdot}72 \qquad 1{\cdot}45$$

for the corresponding total bond-orders. Hence a bond diagram for this excited state of butadiene would be approximately as schematically shown in Fig. 4-4; the middle bond is more like a double bond than the outer ones.

1 2 3 4

FIG. 4-4 Schematic representation of the first excited-state of butadiene.

It is evident, therefore, that in this excited state the butadiene system is highly reactive at its end-points because it is almost like a "double" free-radical.

(c) Bond-Order–Bond-Length Correlations

We have seen from the above discussion that, by means of this bond-order concept, we are beginning to derive information, not just of a qualitative but of a semi-quantitative nature, about the bonds of a conjugated system, both in its ground- and excited states. The greatest advantage of the bond-order index, however, is that we can use it to say something—again, not just qualitative but semi-quantitative—about bond *lengths*.

Let us first consider bond lengths in three different types of carbon-carbon bonds. A single bond is typified by that in saturated paraffin molecules such an ethane. Here the carbon-carbon bond-length is found experimentally to be *ca.* 1.54 Å; as we have seen, the carbon-carbon bond in ethylene may be regarded as the prototype "pure double-bond" and its length, from experimental measurement, is estimated to be *ca.* 1·34 Å. A carbon-carbon linkage in acetylene is *ca.* 1.20 Å in length. Naturally enough, we see the trend "the higher the bond order, the shorter the bond". It was therefore suggested many years ago by Pauling and his collaborators[R8] that a curve ought to be constructed through these three points, called an order-length curve (Fig. 4-5), the pure single-, double- and triple-bonds having, of course,

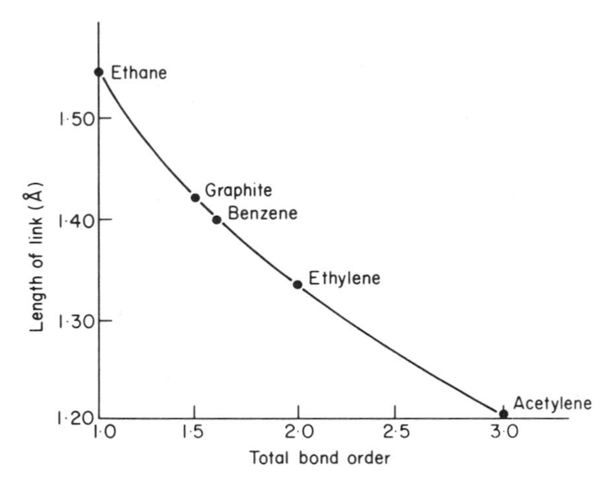

FIG. 4-5 Bond-Length—Bond-Order curve. (Redrawn from: C. A. Coulson's *Valence*, Second Edition, Oxford University Press, London, 1961, p. 270).

total bond-orders of 1, 2 and 3, respectively. The plot obtained is not quite a straight line but is slightly curved. If we could assume that a curve of this kind has a universal validity for bonds (like those in butadiene) whose total bond-orders are not integral—and this is by no means obvious, it would have to be verified—it would be possible to infer the *length* of any given bond, once having calculated its bond order from equation (4-6). We would then expect that, in the unsaturated, conjugated systems of the type we have been discussing in this book, there would be a spread of bond lengths, corresponding to a range of bond orders. This is indeed what is found in practice by means of X-ray crystallography.

The most obvious example on which to test the predictions of this curve is benzene. The Coulson π-bond-order for benzene (which could be calculated from equation (4-6) and the LCAO-coefficients for benzene) is $p^{\pi}_{\text{benzene}} = \frac{2}{3}$ and thus the total bond-order for benzene, $p^{\text{Total}}_{\text{benzene}}$, is *ca.* 1·67. From Fig. 4-5, a bond order of 1·67 corresponds to a bond length of 1·40 Å. The observed value for this bond length in benzene is 1·397 Å and hence this is very satisfactory. We might now therefore further "calibrate" this curve by adding the point (1·67, 1·397 Å) for benzene to the three which already define the plot reproduced in Fig. 4-5. We could, if we wished, include on this curve a few more points which correspond to experimentally well-determined bond-lengths; for example, when this scheme was first developed, graphite was included since it may be thought of as an infinitely extended molecule of this kind.

The question still persists—is this a curve of universal validity? In other words, is it legitimate to do what we have just done for benzene for *any* bond whose order we know and whose length we wish to infer, or was the apparent success with benzene just a fortunate coincidence? The only way to answer this is to take all the experimental results which have been obtained and see if they fit a bond-order–bond-length curve. It turns out that, when this is done, all the available points fall in a narrow band within 0·01 Å either side of the curve shown in Fig. 4-5. There are, of course, errors in the measured bond-lengths and approximations in our Hückel model and hence we would not expect to obtain a perfect experiment-*vs.*-theory fit. The range of uncertainty, 0·01 Å, is thus acceptable, under these circumstances; we can therefore conclude that the idea of a relatively universal carbon-carbon bond-length-*vs.*-bond-order curve has been effectively substantiated. From this it appears that, if required, we would be justified in using such a curve to predict the actual bond-lengths of an arbitrary conjugated-hydrocarbon, from its corresponding HMO-calculated bond-orders. As an example, Fig. 4-6b shows the calculated and observed lengths of the 12 non-equivalent carbon-carbon bonds in the condensed, benzenoid hydrocarbon, ovalene (Fig. 4-6a).

This leads us to the discussion of a very old idea. Older chemists used to talk about "bond fixation". They realised that in benzene it was impossible to "fix" some bonds as double and others as single, as in an isolated Kekulé-structure, because all bonds in benzene are equivalent. They reasoned that this did not mean, however, that it was impossible to fix specific bonds in other, less-symmetrical, molecules such as naphthalene (Fig. 4-7). Much argument took place about where the double bonds in a molecule like naphthalene ought to be placed. They had reasons, based on chemical reactivity, for supposing that the 1-2 bond (and thus also the symmetrically related ones, 3-4, 5-6 and 7-8) were double bonds; this arrangement would of course dictate that the 9-10 bond would also have to be double and thus the structure in Fig. 4-7 became the most favoured one. It used to be asked—are the double bonds really fixed like this? There was in fact no way of concluding whether they were or not until, on the one hand, the experimental technique of X-ray crystallography became available, and, on the other hand, bond-order calculations of the type being described could be performed. With the hindsight gained from these, it is now realised that the "bond-fixation" kind of language is inappropriate; one would no longer say that the double bonds are "fixed" in any one of the possible "Kekulé"-arrangements such as that in Fig. 4-7. What can now be said is that there is a number

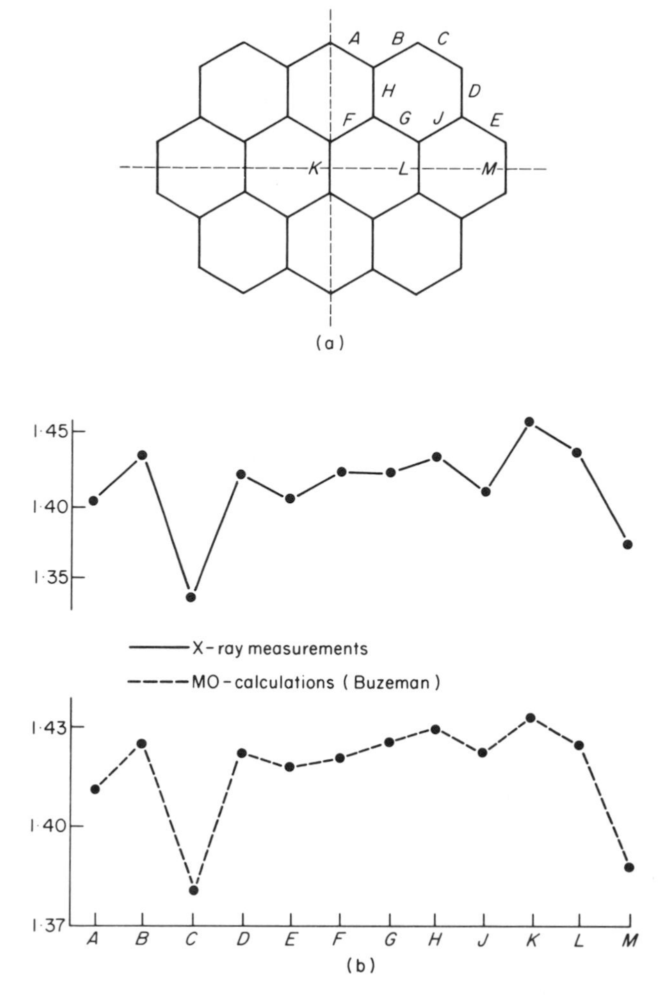

FIG. 4-6 (a) Non-equivalent carbon-carbon bonds (*A* to *H*, *J* to *M*) in ovalene (b) Comparison of calculated and experimentally-observed bond-lengths in ovalene (Redrawn from A. J. Buzeman, *Proc. Phys. Soc. London*, **A63**, 827, (1950).)

FIG. 4-7 A Kekulé form of naphthalene, with central (9-10) bond "double".

to be associated with each bond—its bond order. The total bond-order of the 1-2 bond in naphthalene is *ca.* 1·72 and we can thus say that this bond is more nearly a double bond than its neighbours but that it is of course not *exactly* a double bond. Hence, the old picture of "bond fixation" has been displaced by this description of bonds in terms of their "fractional" (*i.e.*, non-integral) bond-orders.

Let us now consider an example of how these ideas may be applied experimentally; we consider the molecule variously known as acraldehyde or acrolein, depicted in Fig. 4-8. It is a planar system and we can therefore speak

$$\ddot{O}{=\!=}CH{-}CH{=\!=}CH_2$$

FIG. 4-8 Structural formula for acraldehyde.

of some of its electrons as having π-symmetry. There are in fact four π-electrons to consider and hence the system is, in many respects, like butadiene. If we were to insert into the appropriate secular determinant for acraldehyde the Coulomb terms for oxygen from Table 3-1 and perhaps allow resonance integrals to be different from the standard value, β (as *per* Table 3-2), it would be possible to obtain a detailed picture of this molecule. There is an excited state of acraldehyde which can be attained by promoting a non-bonding electron on oxygen (previously not in the π-system) into an anti-bonding π-state; (the oxygen atom was already contributing one electron to the π-system when the molecule was in its ground state). The antibonding π-MO into which this oxygen electron has been excited is of course spread over the framework of the entire molecule. In terms of the butadiene system, (Fig. 2-7) the antibonding π-MO of acraldehyde which receives this excited oxygen-electron would be called Ψ_3. In fact, in terms of the butadiene-set of energy-levels depicted in Fig. 2-7, the excitation may be represented:

$$(\Psi_1)^2(\Psi_2)^2 + 1\text{ e} = (\Psi_1)^2(\Psi_2)^2(\Psi_3)^1 \qquad (4\text{-}14)$$

This process, in which a non-bonding electron around the oxygen atom has been converted into an anti-bonding π-electron, is sometimes referred to as an "$n \rightarrow \pi^*$ transition", and it has been very carefully studied, in the region 3800 Å, by Brand and Williamson[R9]. They succeeded in resolving a rotation vibration spectrum of acraldehyde and, as a result, were able to make some inferences about the changes in bond lengths which occur upon such an $n \rightarrow \pi^*$ excitation.

Let us first consider what sort of changes we might expect. We shall use the butadiene example for the purposes of illustration; even though its LCAO-coefficients will not be exactly the same as those which would be found from a calculation on acraldehyde itself, they will suffice for the purposes of this discussion. Essentially then, to simulate, in butadiene, the $n \rightarrow \pi^*$ transition in acraldehyde, we are going to consider adding an electron to the Ψ_3-orbital of butadiene, without taking anything away from the ground state, $(\Psi_1)^2(\Psi_2)^2$; this results in $(\Psi_1)^2(\Psi_2)^2(\Psi_3)^1$. This process will therefore weaken the end-bonds and strengthen the central one. From this we should expect Δr_{12} (the change in the length of the 1-2 bond during the electron-addition process) to be *positive*—*i.e.*, $\Delta r_{12} > 0$; as for their actual magnitudes, a calculation using these butadiene data suggests that the decrease in length of the central bond would be somewhat less than the accompanying increase in the length of the end-bonds, as the calculated numbers in Table 4-3 indicate.

TABLE 4-3

Bond-Length Changes Consequent upon an $n \rightarrow \pi^*$ Transition in Acraldehyde

	Calculated (Using Butadiene LCAO-MO Data)	Observed (For Acraldehyde)
Δr_{12}	0·04 Å	0·06 Å
Δr_{23}	−0·03Å	−0·04Å

The experimentally observed values in Table 4-3 are not guaranteed to the last decimal place quoted but are *probably* accurate to 0·01 Å; furthermore, we have used the butadiene data to approximate what would have been calculated for acraldehyde. In view of these two factors, the agreement displayed in Table 4-3 is as good as (and maybe even better than!) one has the right to hope for.

In the next section (still in the context of bond orders) we discuss another index which may be calculated from Hückel LCAO-coefficients and which also links with older ideas in chemistry; we refer to *free valence*.

4.4 Free Valence

(a) The Concept and Definition of Free Valence

It is now over 150 years since Thiele introduced the idea of "partial valence" and Werner talked about the "residual affinity" of an atom in a molecule. The limitation of these older concepts was that they were, of necessity, only qualitative. Let us consider again our standard case, butadiene; these older chemists supposed that the bond structure was as in Fig. 4-9, in which the dotted line is intended to denote something weaker than a single bond; (this is quite acceptable, for, from our MO-model, we know that this *is* the situation in butadiene). This therefore left some bonding propensity—what might be

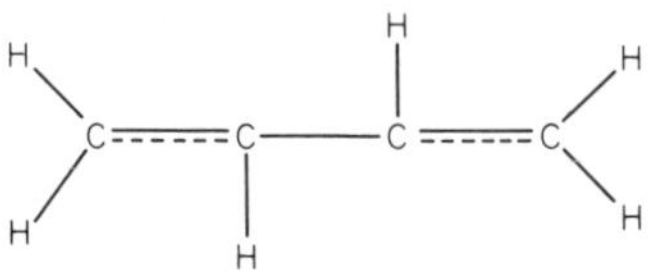

FIG. 4-9 Pre-quantum-mechanical view of the carbon valencies in butadiene.

termed "unused bonding-power"—at the two end-carbon-atoms; hence chemists of that era developed the idea, which had to remain only qualitative, that there should be some "residual affinity" or "partial valence" at the two end-carbon-atoms. (There will also be a certain amount at the two middle ones but of course the old problem of whether butadiene reacted at the (1,4) or (2,3) positions soon became embroiled in that line of discussion). We can now introduce an index, calculable from Coulson bond-orders, which has the great advantage that it can actually give numerical magnitude to the old, qualitative concept of residual affinity; this is the *free valence* index, F_r, for a given atom r in an arbitrary conjugated system.

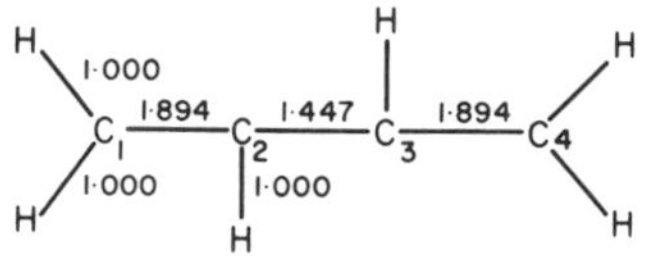

FIG. 4-10 Total bond-orders of carbon-carbon and carbon-hydrogen bonds in butadiene.

Once more, we use butadiene as the example; Fig. 4-10 shows the *total* bond-orders of all bonds in butadiene, including the C—H bonds which normally do not feature in the considerations of a Hückel calculation; (since they are pure, single, σ-bonds, their total bond-order will be, on these assumptions, precisely unity). By examination of a diagram of this sort it would be a trivial matter to add up the total order of all the bonds which terminate on some particular atom, r; let us call this sum the total bond-number for this atom and denote it N_r. For example, it is clear that

$$N_1 = N_4 = 1{\cdot}894 + 2 \times 1{\cdot}000 = 3{\cdot}894 \tag{4-15}$$

In the same way, N_2 $(= N_3)$ can be obtained from the data of Fig. 4-10; thus,

$$N_2 = N_3 = 1{\cdot}894 + 1{\cdot}447 + 1{\cdot}000 = 4{\cdot}341 \tag{4-16}$$

Hence we see that the two non-equivalent carbon-atoms in butadiene have rather different total bond-numbers; the end-carbon-atoms are nothing like so deeply involved in bonding (total bond-number: 3·894) as are the middle atoms (total bond-number: 4·341). In an intuitive and heuristic way, this may be taken to imply that the end-atoms have more potential bonding available for some possible reaction than do the middle ones. This last statement may be put more quantitatively; for it can be shown that there is a maximum value (denoted N_{max}) of the quantity N_r which a given trigonally-hybridised carbon-atom may attain. It is

$$N_{max} = 3 + \sqrt{3} = 4{\cdot}732 \qquad (4\text{-}17)$$

(Parenthetically, we might mention that in the case of a carbon-atom involved in two double bonds (=C=),

$$N_{max} = 2 + 2\sqrt{2} = 4{\cdot}828 \qquad (4\text{-}18)$$

However, in the present discussion we shall be concerned only with trigonal carbon-atoms, for which $N_{max} = 4{\cdot}732$, as in equation (4-17)).

We now know what the *maximum* total bond-number can be for any trigonal carbon-atom (such as the carbon-atoms in butadiene) and we also know what the actual N_r-values are for the various carbon-atoms in butadiene. The difference between these two quantities is what we might regard, bearing in mind the older ideas of Thiele and Werner, as the "unused bonding" or the *free valence*, F_r, of the rth atom. Thus:

$$F_r = N_{max} - N_r \qquad (4\text{-}19)$$

The data of equations (4-15)–(4-17), together with the expression (4-19), lead to

$$\left.\begin{aligned} F_1(=F_4) &= 0{\cdot}84 \\ F_2(=F_3) &= 0{\cdot}39 \end{aligned}\right\} \qquad (4\text{-}20)$$

for butadiene. By this device we have therefore now made quantitative what hitherto had been only qualitative.

(b) Free Valence and Chemical Reactivity

Table 4-4 illustrates some other examples of free valences, at the atoms indicated, in various molecules. It is quite clear that these values are consistent with chemical experience. Benzene is stable, with a small free-valence, ethylene is more reactive; benzyl is a radical and hence able to react very freely, and the quinodimethane species behaves almost like a diradical. This type of analysis is therefore qualitatively sensible and has the further merit of being a semi-quantitative measure of reactivity. For some reactions, it is possible to obtain a very close correlation between the rate at which the reaction takes place and the free valence of particular centres in the reactants.

The free-valence index is actually found to be of greatest importance in free-radical reactions. Early work by Sware[R10] investigated the rate

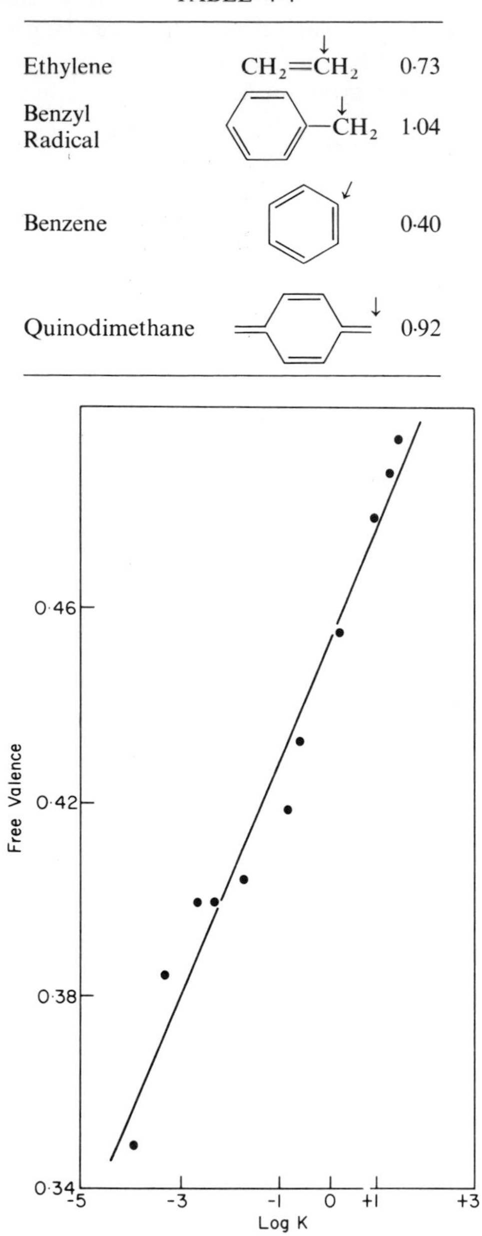

TABLE 4-4

Ethylene	$CH_2{=}\overset{\downarrow}{C}H_2$	0·73
Benzyl Radical	$C_6H_5{-}\overset{\downarrow}{C}H_2$	1·04
Benzene	C_6H_6	0·40
Quinodimethane	$CH_2{=}C_6H_4{=}CH_2$	0·92

FIG. 4-11 Relation between reaction-rate for CCl_3-addition and maximum free-valence, for a series of molecules. (Redrawn from E. C. Kooyman and E. Forenhorst, *Trans. Faraday Soc.*, **49**, 58 (1953).)

at which the —CH_3 radical attacked a wide range of aromatic hydrocarbons, while Kooyman and Fohrenhorst[R11] similarly studied the reactivity of the —CCl_3 radical. In this latter investigation, the logarithm of the rate of reaction was plotted (along the *abscissa*) against calculated free-valence (along the ordinate) at the particular centre in the molecule in question being attacked by the —CCl_3 radical. The data for a set of approximately 25 points (involving —CCl_3 attack on a wide range of conjugated systems from small molecules to those comprising three or four condensed rings) gave rise to such a good straight line (Fig. 4-11) that it would certainly be possible, if required, to use this line to predict the rate of —CCl_3 attack on some new hydrocarbon, once the free-valence indices of its various carbon-atoms had been calculated from equation (4-19). Such a procedure does not, of course, reveal anything about the *mechanism* of such a radical attack—and, indeed, it might be held to have some empirical character about it; nevertheless, it does work very satisfactorily, and a similar correlation is obtained when the —CH_3 radical is the attacking agent; hence this approach is one of fairly wide applicability.

4.5 Concluding Remarks

The three MO-indices discussed in this chapter—charge, bond order and free valence—are calculable solely from a given electronic configuration and the Hückel LCAO-MO-*coefficients* of an arbitrary, planar, conjugated system (hydrocarbon or hetero-molecule). As was emphasised at the end of Chapter Three (§3.5), these quantities are therefore independent of the actual numerical values assumed for the parameters α and β, no mention of which whatsoever, it will be noted, has been made in the current chapter. In the next chapter we shall see what information (likewise not dependent on α and β) can be obtained from the LCAO-MO-*energy levels* of, specifically, monocyclic, planar, conjugated systems.

Five

The Hückel "4p + 2" Rule of Aromaticity

5.1 Introduction

In pre-wave-mechanical days it was noticed by organic and inorganic chemists such as the late Sir Robert Robinson and the late Sir Christopher Ingold that ring systems with six π-electrons enjoyed a peculiar stability, apparently surviving intact a variety of experimental conditions. In order to describe the electronic configuration of these cyclic, six-π-electron-systems they coined the term "aromatic sextet". We shall see in the present chapter how the exceptional stability of the aromatic sextet is rationalised, in a quite natural way, by Hückel theory. We shall further show that the pronounced stability of a six π-electron-system is merely a special case of a more-general rule—which has become known as Hückel's Rule of Aromaticity since it was observed by Hückel in his original series of papers[R12]—which associates particular stability with planar, cyclic, π-electron-systems comprising $(4p + 2)$ electrons, where p is an integer. Putting $p = 1$, we note that $(4 \times 1 + 2) = 6$ and thus realise that the "aromatic sextet" is the first, and certainly the most famous, example of this more-general rule. The stabilities of monocyclic, planar, π-electron systems containing 6, 10, 14, 18, etc. electrons (such as some of the recently synthesised annulenes) are thus intimately bound up with this $4p + 2$ rule.

5.2 The $4p + 2$ Rule

Consider a planar, cyclic, conjugated system C_nH_n, comprising n carbon atoms. Suppose, for the purposes of this discussion, that the Hückel assumptions are valid; we shall examine this supposition in more detail later but for the moment we take it that

a) the systems are planar,
b) we are dealing only with the π-electrons and
c) the carbon-carbon bonds of the system are all equal in length.

We thus neglect any strain which may be caused in the σ-basis as a result of the geometrical constraints just mentioned. Under these circumstances, the σ-bond-carbon-carbon connectivity of the annulene C_nH_n may be represented, for the purposes of a Hückel calculation, by a regular n-gon as in Fig. 5-1,

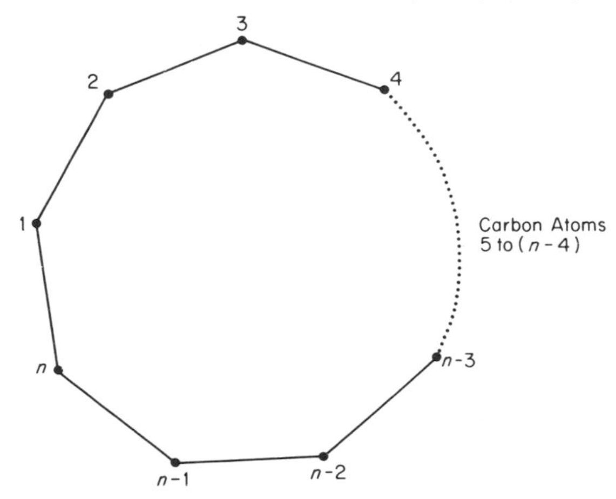

FIG. 5-1 The graph representing the carbon-atom connectivity of an n-annulene, C_nH_n.

and the associated secular determinant is of the form:

$$\begin{vmatrix}
-x & 1 & 0 & 0 & 0 & 0 & \cdots & 0 & 0 & 0 & 0 & 0 & 1 \\
1 & -x & 1 & 0 & 0 & 0 & \cdots & 0 & 0 & 0 & 0 & 0 & 0 \\
0 & 1 & -x & 1 & 0 & 0 & \cdots & 0 & 0 & 0 & 0 & 0 & 0 \\
0 & 0 & 1 & -x & 1 & 0 & \cdots & 0 & 0 & 0 & 0 & 0 & 0 \\
\cdots & \cdots & \cdots & \cdots & \cdots & \cdots & \cdots & \cdots & \cdots & \cdots & \cdots & \cdots & \cdots \\
0 & 0 & 0 & 0 & 0 & 0 & \cdots & -x & 1 & 0 & 0 & 0 & 0 \\
0 & 0 & 0 & 0 & 0 & 0 & \cdots & 1 & -x & 1 & 0 & 0 & 0 \\
0 & 0 & 0 & 0 & 0 & 0 & \cdots & 0 & 1 & -x & 1 & 0 & 0 \\
0 & 0 & 0 & 0 & 0 & 0 & \cdots & 0 & 0 & 1 & -x & 1 & 0 \\
0 & 0 & 0 & 0 & 0 & 0 & \cdots & 0 & 0 & 0 & 1 & -x & 1 \\
1 & 0 & 0 & 0 & 0 & 0 & \cdots & 0 & 0 & 0 & 0 & 1 & -x
\end{vmatrix} = 0$$

By group-theoretical, or graph-theoretical,[N19] methods, this equation can be shown to have n solutions of the form

$$x_k = 2\cos\left(\frac{2k\pi}{n}\right) \tag{5-1}$$

and thus

$$\varepsilon_k = \alpha + 2\beta \cos\left(\frac{2k\pi}{n}\right) \tag{5-2}$$

where the index k successively takes the value[N19] 1, 2, ..., n (or, what is equivalent, 0, 1, 2 ··· $(n - 1)$). Notice that we can predict immediately when non-bonding orbitals (*i.e.*, those for which ε_k of (5-2) is equal to α) are going to arise in molecules of this sort. They will occur only if there exist particular values of k in the sequence $1, 2, \ldots, k, \ldots, n$ (call them p and q) which satisfy

$$\frac{2p\pi}{n} = \frac{\pi}{2} \tag{5-3a}$$

and

$$\frac{2q\pi}{n} = \frac{3\pi}{2} \tag{5-3b}$$

From equation (5.3a) we deduce immediately the condition

$$n = 4p \tag{5-4}$$

while equation (5-3b) leads to

$$q = 3\left(\frac{n}{4}\right) = 3p \tag{5-5}$$

The required condition (5-3a) is the stronger for, *if* there exists a number $p = n/4$ which is integral then it follows necessarily that there shall exist $q = 3(n/4)$ amongst the integers $(n/4 + 1)$, $(n/4 + 2)$, ..., n. We thus conclude that non-bonding orbitals will arise when n, in C_nH_n, is divisible by 4 and, furthermore, such non-bonding orbitals, when they do occur, will do so in *degenerate pairs*. Without any further analysis, therefore, we can expect the possibility of triplet ground-states and thus instability in the annulenes C_nH_n in which $n = 4p$, p integral.

We may glean one further piece of information from equation (5-2) before we proceed to apply it systematically to a series of annulenes, C_nH_n, with specific values of n. For *every* value of n (whether it be odd or even, divisible by 4 or not) there will be an orbital of energy $\alpha + 2\beta\cos(2n\pi/n) = \alpha + 2\beta$, which arises when the running suffix, k, takes on its final value, n. Thus, every annulene, whatever its size, will have a bonding orbital (it will in fact be the *lowest* bonding-orbital) of energy $\alpha + 2\beta$. If n is even (*i.e.*, if it is divisible by 2) there will be some stage in the sequence $1, 2, 3, \ldots, k, \ldots, n$ when the running index k takes on the value $n/2$. For this value of k, equation (5-2) tells us that there will be an orbital (this time it is an anti-bonding orbital) of energy $\alpha + 2\beta \cos(2(n/2)\pi/n) = \alpha - 2\beta$. Thus, for n even, there is again this pairing, around α, of the lowest bonding-orbital with the highest anti-bonding-orbital. In fact, closer study of the values which $\cos(2k\pi/n)$ may take on, when n is even, as k passes through $1, 2, 3, \ldots, n$, reveals that every bonding-orbital is symmetrically paired, about α, with a corresponding anti-bonding-orbital. This is yet one more example of a very general theorem —the Coulson–Rushbrooke Pairing-Theorem—which was also illustrated by the energy-level pattern found for butadiene and which will be discussed and proved in Chapter Six.

As an illustration of these ideas, and of the $4p + 2$ rule, let us consider the energy-level patterns, given by equation (5-2), in the annulenes, C_nH_n, where n runs from 3 to 8. These are shown in Fig. 5-2 in which the energy-level

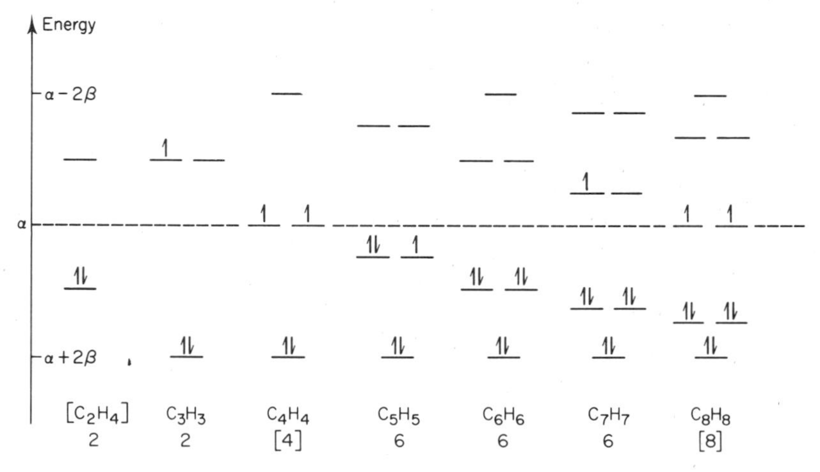

FIG. 5-2 π-Electron energy-levels in ethylene and in the n-annulenes, $n = 3, 4, \ldots, 8$.

pattern of (linear) ethylene is included for comparison[N20]. Let us now consider what is the likely behaviour of systems of this kind.

C_2H_4: We may regard the isolated double-bond as a prototype of this kind of molecule and it is clear from Fig. 5-2 that the two electrons in its π-system fit nicely into the non-bonding orbital available; this bonding orbital is completely occupied while the one non-bonding orbital is empty and this results, as we have seen, in good, solid bonding. Removal of an electron would clearly weaken the bonding since it is a bonding electron which would have to be removed. If we were to try to add an electron to form $C_2H_4^-$ then this would necessitate half-filling the *anti*-bonding orbital which would lead to destabilisation of the resulting species. C_2H_4 is, therefore, most stable as the neutral entity, with its one bonding orbital completely filled and its one

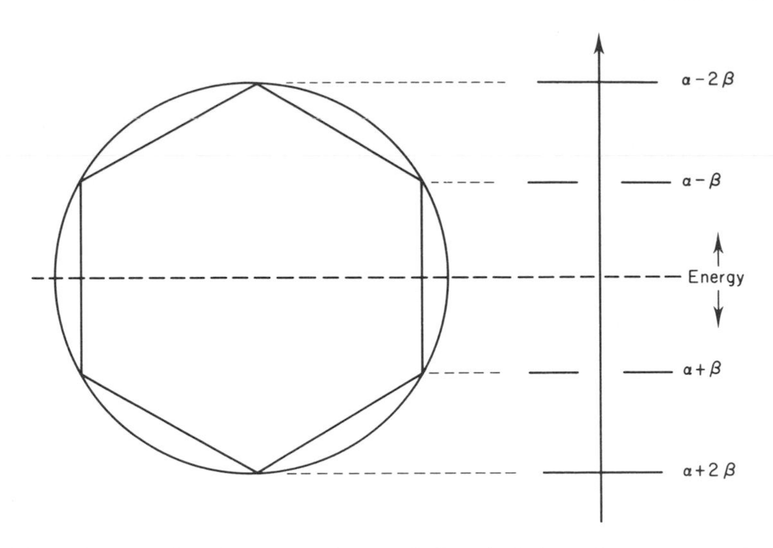

FIG. 5-2a The Frost–Muslin construction applied to find the π-electron energy-levels of benzene.[N20]

anti-bonding orbital unoccupied; this neutral species is therefore more stable than either $C_2H_4^+$ or $C_2H_4^-$.

C_3H_3: In this neutral species there are 3 π-electrons to dispose of amongst the energy levels shown in Fig. 5-2; there is one low-lying bonding-orbital which can readily accommodate two of these, but in neutral C_3H_3 it would be necessary to assign the third electron (with either "spin up" or "spin down", it does not matter which) to one of the two, degenerate, anti-bonding orbitals, as shown in Fig. 5-2. This would not be a very favourable situation, energetically. We should therefore expect that it would be very easy for the system to lose this third (destabilising) electron so that the three-membered-ring system, C_3H_3, would be expected to be more stable (as $C_3H_3^+$) with just *two* π-electrons than with three. Hence, underneath the energy-level diagram for C_3H_3 in Fig. 5-2 is written the figure "2", this being the number of π-electrons we should expect in a system of this sort. Of course, for ethylene this number was 2; we now also have 2 for the three-ring.

C_4H_4: We shall by-pass the C_4H_4-case for a moment; as mentioned earlier, difficulties are to be expected here because of the anticipated presence (confirmed by the energy-level pattern of Fig. 5-2) of non-bonding orbitals giving rise to an electronic configuration involving unpaired (parallel) spins—a triplet state.

C_5H_5: In the C_5H_5-case, Fig. 5-2 shows that the three bonding-orbitals available can quite adequately accommodate the five π-electrons to be assigned to this energy-level system, without the need to invoke use of the anti-bonding orbitals. Indeed, there is "room" for one more electron in the bonding orbitals; this means that the system, C_5H_5, would be *more* stable as the anionic species, $C_5H_5^-$, since the acquisition of this further electron would cause its bonding orbitals to be completely filled—and not just partially filled, as they are in C_5H_5. Thus we see that the stable number of electrons in this system is 6.

C_6H_6: This is the classic benzene example which we have already dealt with earlier. The three low-lying bonding-orbitals are completely filled, in pairs, by the six π-electrons which need to be accommodated in the neutral system. This is truly the "aromatic sextet"!

C_7H_7: In this cycloheptatrienyl species the first six electrons go into bonding levels (Fig. 5-2) and the seventh one has to go into an anti-bonding level. It is not surprising, therefore, that this last electron can easily be removed to form the cation, $C_7H_7^+$, which again has six as the stable number of π-electrons completely filling the bonding-MO's of the system.

Now let us consider the C_nH_n-cases when n is divisible by 4 (i.e., $n = 4$ and 8, of the ones dealt with here) and which, by previous arguments, involve non-bonding orbitals.

C_4H_4: According to the energy-level pattern of Fig. 5-2, two of the four π-electrons of the neutral system would be assigned, without any difficulties, to the lowest (bonding) level, $\alpha + 2\beta$. However, problems do arise in the assignment of the remaining two electrons. For, if we accept the legitimacy of the model we have been using in which all resonance integrals between σ-bonded pairs of atoms involved in the conjugation are the same (and in which, therefore, presumably all bond-lengths are equal), these two remaining electrons have to be distributed amongst *two, degenerate, non-bonding orbitals*. From our previous agreement that this assignment must be carried out in accordance with Hund's Rules of Maximum Multiplicity, as in the *Aufbau* scheme, these two electrons must be assigned to different orbitals with parallel spins. So *if* cyclobutadiene really were a regular, planar, *square*-system then we should expect a triplet to be the configuration of lowest energy. This does not mean to say, however, that butadiene really is a planar, square system; in fact, there is one effect (at least!) which we have neglected. Renner's extension of the Jahn–Teller theorem tells us that in a spin-system of this sort the molecule will distort[R13] in order to remove degeneracies. As a result of this distortion butadiene takes on what can better be described as a rectangular structure (*e.g.*, Fig. 5-3) in which alternate bonds are shorter (*i.e.*, more like double bonds) and the other alternate bonds are longer (*i.e.*, more like single bonds) than the standard carbon-carbon bond-lengths we

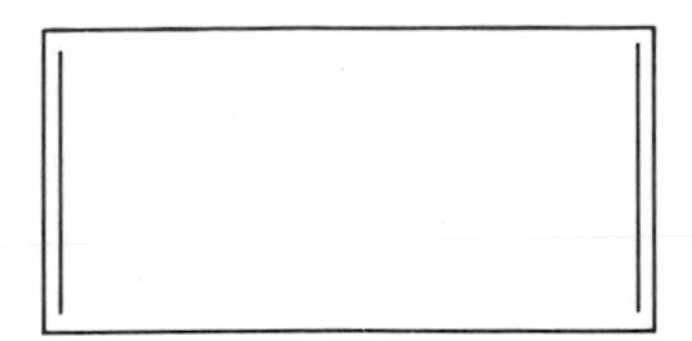

FIG. 5-3 Structural formula of a distorted, "rectangular" form of cyclobutadiene (C_4H_4) which arises because of a pseudo Jahn–Teller or, more strictly, a Renner effect.

have been considering up to now in the regular, square structure for butadiene. A conjugated system with such alternating bond-lengths is of course characterised by *two* resonance integrals: let us call them β_l and β_s for the "long" and "short" bonds, respectively; then $|\beta_l| < |\beta_s|$. It turns out that when this *bond alternation*[N21] (as it is called) is taken into account, the energy-level pattern for butadiene is as in Fig. 5-4b instead of that in Fig. 5-4a which arises in the case we have been considering until now, in which $\beta_l = \beta_s = \beta$. (Of course, as $\beta_l \rightarrow \beta_s \rightarrow \beta$, the energy-level pattern in Fig. 5-4b collapses to that in Fig. 5-4a).

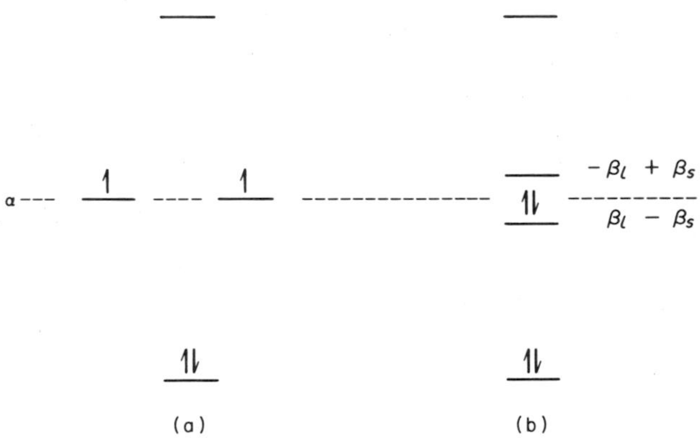

FIG. 5-4 π-Electron energy-levels of cyclobutadiene (a) in its "ideal", symmetrical form, in which all carbon-carbon bond-lengths are equal, and (b) in its rectangular form, caused by a pseudo Jahn–Teller or Renner distortion.

Under these circumstances, a singlet state with paired spins appears to be the ground state of the π-electron system; but cyclobutadiene is, nevertheless, a very unstable species for it must be remembered that in the present approach we have completely neglected any instability due to the necessary geometrical constraints on the σ-hybrids which are involved in maintaining this rectangular configuration of the four carbon-atoms.

C_8H_8: With cyclo-octatetraene we have precisely the same situation, formally, as with C_4H_4 when the simplest HMO-approximations are invoked and the carbon atoms of this molecule are assumed to lie in a plane at the apices of a regular octagon (Fig. 5-2). However, although bond alternation would set in (by the Jahn–Teller arguments outlined above) if the system were approximately planar, in this molecule something far more drastic and serious happens; the strain on the σ-system (which, as has often been emphasised, we are neglecting in this approach) is so severe in the planar conformation that the molecule buckles completely out of plane. This situation is represented schematically in Fig. 5-5. The geometry

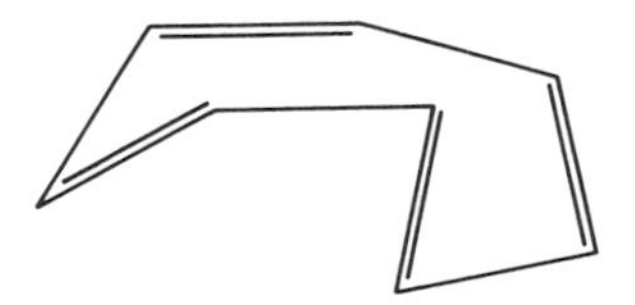

FIG. 5-5 "Buckled", non-planar form of cyclo-octatetraene.

has thus changed completely and although, in this configuration, the eight electrons we are considering (which we cannot really now call π-electrons since there is no longer a molecular plane—see §2.4) do go into bonding orbitals, it is clear that the Hückel picture can no longer apply.

5.3 Extension of the $4p + 2$ Rule

We see, therefore, from the arguments in §5-2, that although there are difficulties with the simple picture in the case of monocyclic hydrocarbons, C_nH_n, when $n = 4p$, the basis of the aromatic-sextet rule has been well rationalised. Benzene is stable as a neutral species, the cyclopentadienyl system is more stable as an anion, $(C_5H_5)^-$, and the cycloheptatrienyl system is more stable as a cation, $(C_7H_7)^+$, and these observations are all well substantiated experimentally. We now attempt to extend the rule, in a heuristic way, to systems other than monocyclic hydrocarbons and consider some examples of such extensions.

(a) Azulene

This hydrocarbon, $C_{10}H_8$, famous for the blue colour obtained from it, consists of a five-membered ring and a seven-membered ring fused together (Fig. 5-6). Extending our arguments concerning monocyclic hydrocarbon-systems might lead us to suppose that the five-membered ring would be more stable if it had associated with it an additional electron; hence we put

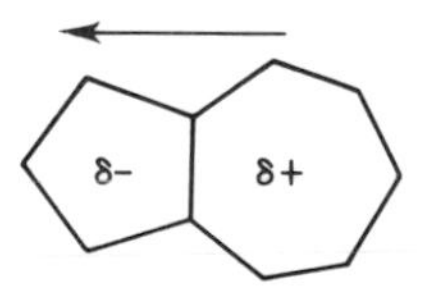

FIG. 5-6 Carbon-atom σ-bond connectivity, and direction of π-electron charge-displacement, in azulene.

a δ- sign on this ring in Fig. 5-6 indicating that this ring will attempt to draw π-electron density onto one or more of the five carbon-atoms in it. We might similarly argue that, on the right-hand side of Fig. 5-6, there is locally a seven-membered ring. A seven-membered ring, according to the reasoning of §5-2, will tend to give up electrons; this ring is therefore prepared to abandon at least part of the density of one of its π-electrons and hence we label the seven-membered ring of azulene in Fig. 5-6 $\delta+$. Immediately, therefore, a dipole moment will be predicted for this molecule, for there is a partial separation of charge within it (the movement of charge to achieve this separation being indicated by the direction of the arrow in Fig. 5-6). Experimentally, there *is* a dipole moment of *ca.* 0·755 Debye units and so, although the arguments put forward here must not be pressed too strongly (indeed they do not bear too close an examination!), we do have, on this simple picture, at least the beginnings of a qualitative explanation for the observed dipole-moment of azulene, which will be substantiated more quantitatively later (§6-5).

(b) Cyclotriapentafulvalene

The second example which we may discuss is a fascinating molecule called cyclotriapentafulvalene which has a three-membered ring joined to a five-membered ring, as in Fig. 5-7. We may argue that the three-membered

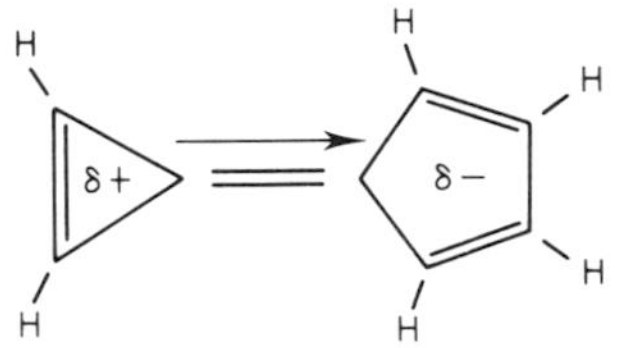

FIG. 5-7 Structural formula of, and direction of π-electron charge-displacement in, cyclotriapentafulvalene.

ring is more stable as a two-electron system ($4p + 2$ with $p = 0$) instead of an approximately three-electron system which we consider it to be initially; hence this ring is labelled $\delta+$; and by the same reasoning as was applied above to the five-membered ring of azulene, the five-membered ring of cyclotriapentafulvalene is denoted $\delta-$. Again, therefore, we predict a large dipole-moment for this molecule. The parent molecule has not yet been synthesised but the hexa-phenyl derivative has; in this derivative the peripheral hydrogen atoms in Fig. 5-7 are replaced by phenyl groups. They will make very little difference to the net result of a dipole-moment measurement on this molecule and for steric reasons they will of course be almost at right-angles to the plane of the cyclotriapentafulvalene skeleton. Now the quite extraordinary finding of an experimental dipole-moment measurement on this molecule is that the dipole moment is no less than *ca.* 6·0 *D*! Here is a hydrocarbon—a pure hydrocarbon, nothing else—and yet it has such an enormous dipole moment of this magnitude. So we see that even though there may be many refinements which we can add to this simple approach, we have at any rate obtained a good deal of insight into the, sometimes-quite-spectacular, properties of these large hydrocarbon-systems.

(c) Pyridine

We need not, however, limit ourselves to (cyclic) hydrocarbons, in our application of the $4p + 2$ rule and its extensions. Let us, for example, consider pyridine (Fig. 5-8). There are six π-electrons and thus six MO's

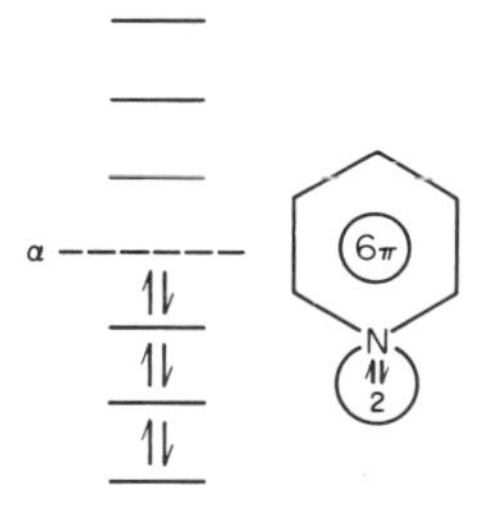

FIG. 5-8 Ground-state, π-electronic configuration of pyridine.

as before, with benzene; they will, furthermore, be related to the MO's for benzene, for one can think of the change from benzene to pyridine as simply involving an alteration in one of the atoms of benzene (see §3.4). In setting up the secular determinant, therefore, we could begin as if the problem were benzene and then make an appropriate change which would take account of the fact that one of the atoms in the conjugated system was a nitrogen atom instead of a carbon. In other words, one could treat the problem of pyridine as a *perturbation* of the benzene problem; there is, in fact, a whole way of doing calculations of this sort in which one goes from the hydrocarbon to the hetero-conjugated system by means of a series of small perturbations.

Alternatively, the secular determinant could be solved directly as described in §3.4. Whatever method is employed, the energy-level pattern which is obtained in the case of pyridine differs from that of benzene in that the two degeneracies which are evident in the MO-energies of benzene are split in the corresponding energy-level pattern for pyridine, as Fig. 5-8 shows schematically. Of course, if nitrogen and carbon were nearly equal in electronegativity, then the splitting of the two (originally degenerate) orbitals would be quite small; but since these two atoms do in fact differ greatly in electronegativity this splitting is quite large. Note also, in Fig. 5-8, that although there are still three bonding-orbitals and three anti-bonding orbitals these are now no longer paired symmetrically about α as was the case for benzene. However, since there *are* three bonding-orbitals these may all be completely filled by six π-electrons and so six is again the stable number of electrons for this system. Furthermore, six is precisely the number of electrons in the π-system of pyridine, which must be accommodated in the orbitals shown in Fig. 5-8. This is so because, as was argued in §3.3a, the nitrogen atom in pyridine, like the carbon atoms, effectively contributes one electron to the π-system. There are thus six π-electrons in all and hence we should not be surprised at the overall stability of pyridine.

(d) Pyrrole

By a similar process it is found that in pyrrole (Fig. 5-9) there are three bonding-orbitals and two anti-bonding ones. Once more, the presence of three bonding-orbitals means that a configuration of six π-electrons will be the most stable. As argued in §3.3a, the nitrogen atom in pyridine may be considered to donate *two* electrons to the π-system; since each of the four carbon-atoms provides one π-electron this gives, once more, a total of six π-electrons and thus a stable entity.

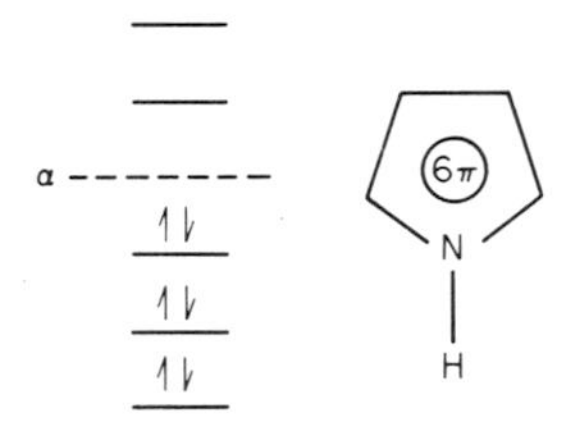

FIG. 5-9 Ground-state, π-electronic configuration of pyrrole.

(e) Larger Monocyclic Systems: *e.g.* [18]-annulene

As we have seen in the case of C_8H_8, some of the higher members of the series C_nH_n are under considerable steric strain if they attempt to assume

120° valence-angles *and* stay planar. The most famous of the larger annulenes (as they are called) which appears to exist comfortably in the planar conformation is [18]-annulene (Fig. 5-10) which is an example of the $4p + 2$

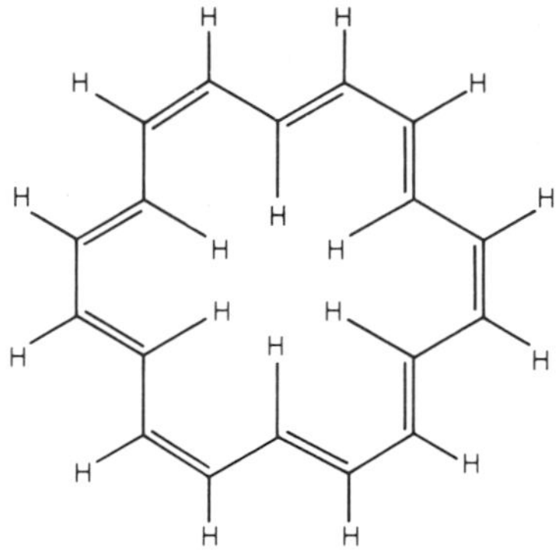

FIG. 5-10 Structural formula of [18]-annulene.

rule, with $p = 4$. The 120° valence-angle arrangement can be achieved, not by having the carbon atoms situated on the vertices of a regular polygon with 18 sides, but by having them situated approximately as the carbon atoms on the periphery of the stable and well-characterised, condensed, benzenoid hydrocarbon, coronene (Fig. 5-11). This arrangement divides the 18 hydrogen

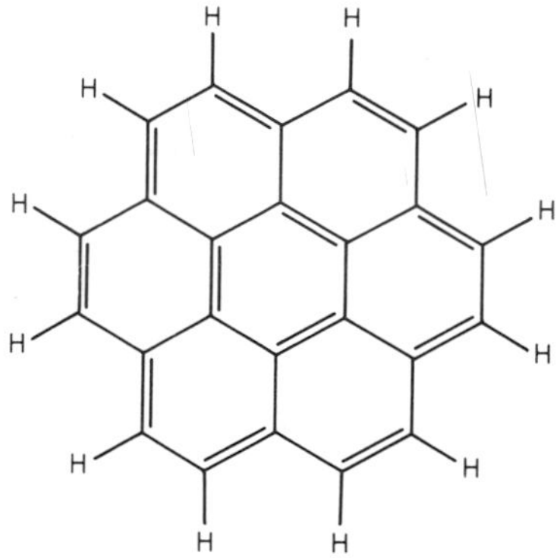

FIG. 5-11 Structural formula of coronene.

atoms of the annulene into two classes: 12 (all equivalent) "external" ones on the periphery and 6 (all equivalent) "internal" ones. The experimental proton-magnetic-resonance spectrum of [18]-annulene consists of two peaks, one at very high field corresponding to the 6 inner protons, and one (of double the intensity) at very low field, corresponding to the twelve outer protons—entirely in accord with the diamagnetic character which HMO-theory predicts a $(4p + 2)$-annulene to have. For $4p$-annulenes, HMO-theory predicts[R14] the *inner* protons to resonate at low field and the *outer* ones to resonate at high field in a proton-magnetic-resonance experiment—exactly the reverse of the situation which obtains with the $(4p + 2)$-annulenes—since the π-electron systems of $4p$-annulenes are predicted by HMO-theory to be *para*magnetic. It has been one of the major triumphs of HMO-theory in recent years that this unambiguous and clear-cut prediction has been dramatically and spectacularly verified by experimental measurements.[R14]

(f) Polycyclic Systems

We have so far applied the $4p + 2$ rule only to *mono*cyclic systems of various types; this is quite proper since the rule has been devised formally only for monocycles (§5.2). However, attempts have been made, in a heuristic and unsubstantiated way, to apply a modified form of the rule to polycyclic systems. This is done by considering only the *peripheries* of the molecules in question, any cross-links which convert the otherwise-monocyclic periphery into part of the polycyclic system being considered as small perturbations. For example, naphthalene, with its 10-π-electron system ($10 = 4p + 2$, with $p = 2$), may be considered as a perturbed [10]-annulene and is thus stable (Fig. 5-12a).

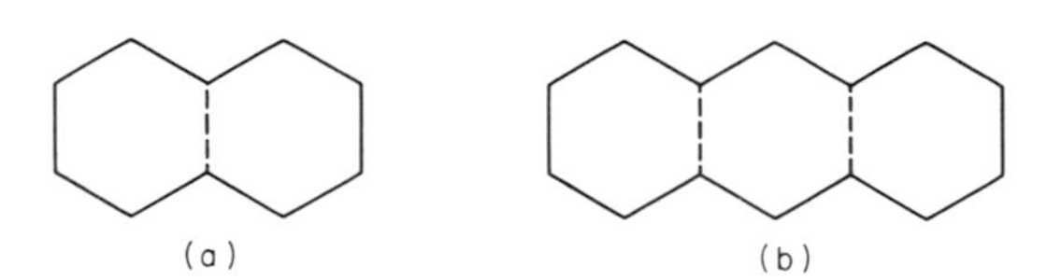

FIG. 5-12 (a) Naphthalene and (b) anthracene considered as "perturbed" $(4p + 2)$-annulenes.

The argument could also be applied, for example, to anthracene (Fig. 5-12b) which has a 14-atom periphery and is also therefore a $(4p + 2)$-(now with $p = 3$) π-electron system which is thus stable. Recently, this type of reasoning has been stretched even further by the argument that pyracylene, for example, (Fig. 5-13) although in fact a 14 π-electron system, may be considered as a "perturbed [12]-annulene" because its periphery involves 12 carbon-atoms.[R15] It has been argued therefore that the π-electron system of this molecule should display the paramagnetic behaviour characteristic of the $4p$-annulenes, even though the molecule is actually, in total, a $(4p + 2)$-π-electron system.[R15] It cannot be denied that such arguments do rationalise

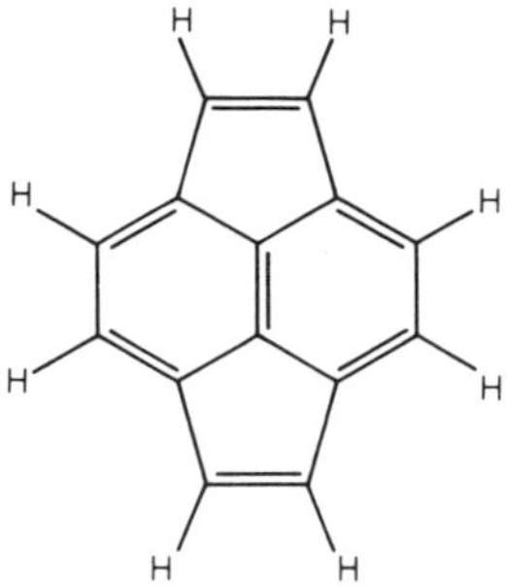

FIG. 5-13 Structural formula of pyracylene.

the observed proton-magnetic-resonance spectrum of pyracylene, but the *a priori* prediction of the overall diamagnetic or paramagnetic nature of *poly*cyclic hydrocarbons, merely by inspection, on the basis of their carbon-atom connectivities, remains a somewhat hazardous undertaking.[R16]

Six

Alternant and Non-Alternant Hydrocarbons: The Coulson–Rushbrooke Theorem

6.1 Classification of Alternant Hydrocarbons: the "Starring" Process

In this chapter we shall make the fundamental distinction between *alternant* and *non-alternant* hydrocarbons and investigate some of the general properties of the MO energy-levels and LCAO-MO coefficients of the former class of hydrocarbons. We shall approach this in a rather formal, mathematical way; this is entirely appropriate since the π-electron energy-levels and wave-functions of alternant hydrocarbons are intimately bound up with the eigenvalues and eigenvectors, respectively, of certain real-symmetric matrices—in this case they are the Hamiltonian matrices (equation (2-30))—which may be partitioned in a particular way. (This is explained in Appendix D to which the more mathematically-inclined reader may refer *after* reading the present chapter). We restrict ourselves, in this discussion, to hydrocarbons only, since, although the classification can be extended in various ways to hetero-conjugated systems, it is in the field of hydrocarbons that it has had its greatest successes and its most useful applications.

We first define a process which has become known as the *starring process*[N22]. The starring process asks us to consider the carbon-atom connectivity of a conjugated hydrocarbon-system (such as, for example, that of naphthalene—Fig. 6-1a) and to place a *star* on certain of the carbon-vertices, leaving the others unmarked in order that we may refer to "starred" and "unstarred" atoms. The starring process must be such that, when it is completed, all the atoms have been assigned to the two groups (starred and unstarred) in such a way that no member of one group is adjacent to (*i.e.*, on the σ-framework, is bonded to) another atom of that group; in other words, we never find two atoms of the same kind, starred or unstarred, as nearest neighbours.

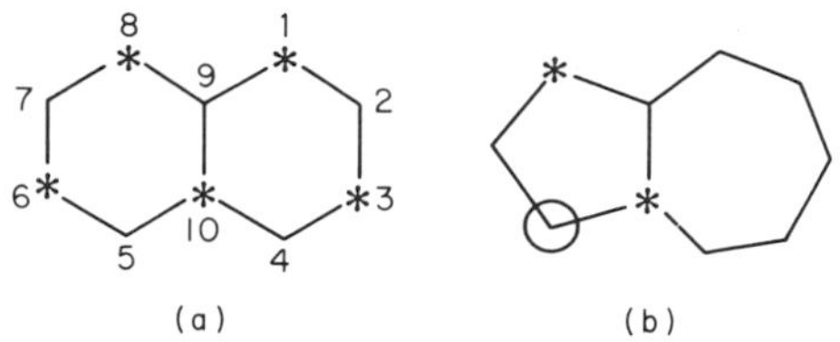

(a) (b)

FIG. 6-1 The "starring process" applied (a) successfully to naphthalene, (b) unsuccessfully to azulene.

Suppose, for example, that, in naphthalene, we start by arbitrarily assigning a star to the 1-position (Fig. 6-1a); then the 2-position must be left unstarred, the 3-position will be starred, the 4-position will be unstarred. Similarly, the atom in the 10-position would have to be starred (because it is next to an unstarred one, in position 4) and that would require the 9-position and the 5-position to be unstarred, etc. This process can then be continued, in similar vein, around the second ring thus showing us that the starring process is perfectly possible for naphthalene. There are molecules, however, for which this arrangement cannot be achieved; let us consider the isomer of naphthalene, azulene (Fig. 6-1b). Here the starring process cannot be carried out because the carbon atom indicated by a circle in Fig. 6-1b would have to be unstarred since it is next to a starred atom and, at the same time, it would have to be starred since it is also adjacent to a neighbouring unstarred-atom! For this molecule, the "starring process", as defined, is thus impossible.

Those hydrocarbon systems whose carbon-atom connectivities allow the starring process to be carried out are classified as *Alternant Hydrocarbons* (AH or "Alternants"): those for which no starring process is possible are classified as *Non-Alternant Hydrocarbons* (NAH or, more colloquially, "Non-Alternants"). A little thought will convince the reader that the necessary and sufficient condition for a hydrocarbon to be alternant is that it shall contain *no odd-membered rings.* All conjugated hydrocarbons which are acyclic (such as butadiene) or which consist solely of condensed, benzenoid rings (such as naphthalene, anthracene, and coronene (Fig. 5-11), for example) are thus alternant. Those containing odd-membered rings, such as azulene, fluoranthene (Fig. 6-2) and the *odd*(*n*)-annulenes in Fig. 5-21, are all non-alternant. The classification applies not only to molecules but extends to radicals—*e.g.* the benzyl radical (Fig. 6-3) is alternant.

In this case there happen to be more starred vertices than unstarred. This is of no consequence; indeed, there are systems in which the difference between the number of atoms in the two groups is greater than one. Although it is immaterial which of the two distinct groups of vertices is arbitrarily called "starred" and which "unstarred", by convention the larger group of vertices is called the "starred" set. We now consider an important consequence of this alternant/non-alternant classification.

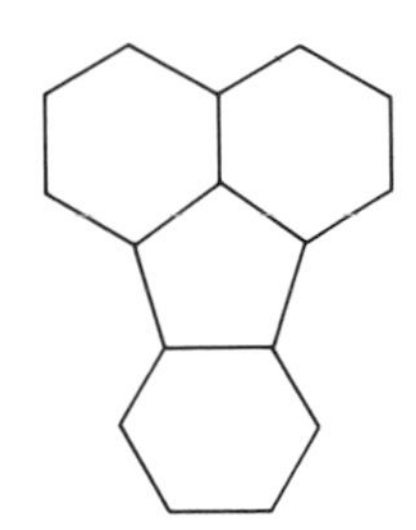

FIG. 6-2 The carbon-atom connectivity of the non-alternant hydrocarbon, fluoranthene.

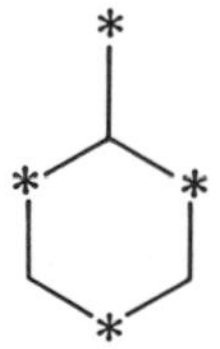

FIG. 6-3 The "starring process" applied to the benzyl radical.

6.2 Statement of the Coulson–Rushbrooke Theorem

This famous three-part theorem, which constitutes one of the basic foundations of Hückel theory, may be stated as follows:

1) "In an alternant hydrocarbon the MO energy-levels are symmetrically paired about an appropriate zero (α) such that if $\varepsilon_I = \alpha + k\beta$ is a root of the secular equations then $\varepsilon_{n-I+1} = \alpha - k\beta$ is also a root; there is thus a *complementary character* about the molecular-orbital energies".

A more formal way of stating this is to say that, for alternant hydrocarbons,

$$\varepsilon_I + \varepsilon_{n-I+1} = 2\alpha \tag{6-1}$$

for all I, $1 \leq I \leq n$.

This type of energy-level pattern was evident in the MO energy-spectrum of the alternant hydrocarbon, butadiene, discussed in Chapter Two (§2.7) and in the energy levels of the [n]-annulenes (n even) of Fig. 5-2.

2) "In an alternant hydrocarbon, the LCAO-coefficients of any pair of complementary orbitals (*i.e.* orbitals of energy $\alpha + k\beta$ and $\alpha - k\beta$) are *identical apart from a change of sign in the coefficients of the atomic orbitals centred on the unstarred atoms*".

An illustration of this aspect of the Coulson–Rushbrooke Theorem is again provided by the alternant hydrocarbon, butadiene. The LCAO-coefficients of the two pairs of complementary orbitals in the molecule display this alternation of sign, as examination of equations (2-67) confirms; (atoms 1 and 3 of Fig. 2-6 may be considered, for this purpose, as the "starred" atoms).

3) "In a neutral, alternant hydrocarbon or hydrocarbon-radical, atomic charges on the various carbon-atoms which are involved in the conjugation are, in the ground state, *all precisely unity*".

The remaining sections of this chapter are devoted to formal proofs of these three distinct parts of the Coulson–Rushbrooke Theorem. There are very many and varied (though, of course, equivalent) proofs of part 1 of the theorem; the particular exposition detailed in the next section (§6.3) will not necessarily be the one which the reader will find the most obvious or straightforward. It does, however, have a certain aesthetic charm which should become evident as the proof progresses!

6.3 Proof of Part 1 of the Theorem (Concerning Pairing of the Energy Levels)

We must find first of all a convenient way of labelling the atoms of a given conjugated hydrocarbon. For purely "chemical" purposes the atoms are normally numbered in sequence around rings, etc. However, this is not the most convenient way for the proof which concerns us at present. We shall choose[N23] to call the "starred" atoms (let there be m of them) $1, 2, 3, \ldots, m$ and the "unstarred" atoms $m+1, m+2, \ldots, n$; (as before we consider there to be n carbon atoms in the conjugated system). Note that we are not presuming that there are necessarily the same number of starred and unstarred atoms; if there were the same number in both groups, then m would be equal to $n/2$, but in the present discussion we are considering the more general case.

In what follows we make, to begin with, the simple Hückel-assumptions—*i.e.*

$$\text{a)}\quad \alpha_r = \alpha, \qquad r = 1, 2, \ldots, m, \ldots, n \tag{6-2}$$

$$\text{b)}\quad \beta_{rs} = 0 \quad (r \neq s)r, s \textit{ non}\text{-neighbours} \tag{6-3}$$

$$\text{c)}\quad \beta_{rs} = \beta \quad (r \neq s), \text{ for all } r, s \text{ adjacent} \tag{6-4}$$

(This latter assumption is not actually necessary for the Pairing Theorem to hold but for simplicity we shall make the assumption here and examine it later.)

$$\text{d)}\quad S_{rs} = \delta_{rs} \tag{6-5}$$

The rth secular equation is

$$(H_{rr} - \varepsilon)c_r + \sum_{s \neq r} H_{rs}c_s = 0 \tag{6-6}$$

With the assumptions outlined above, the H_{rr}-terms in (6-6) are all α, and the H_{rs}-terms are either β or zero. Now, if atom r happens to be a starred atom then all the atoms next to it will be unstarred (by hypothesis, since the system in question is alternant). Consequently, the only H_{rs}-terms which can possibly be non-zero are those for which $s = m + 1, m + 2, \ldots, n$; and if atom r were unstarred, the only H_{rs}-terms which would have a chance of being non-zero would be those for which $s = 1, 2, \ldots, m$. When we come to eliminate the $\{c_r\}$-coefficients we therefore obtain a secular determinant which can be partitioned as follows:

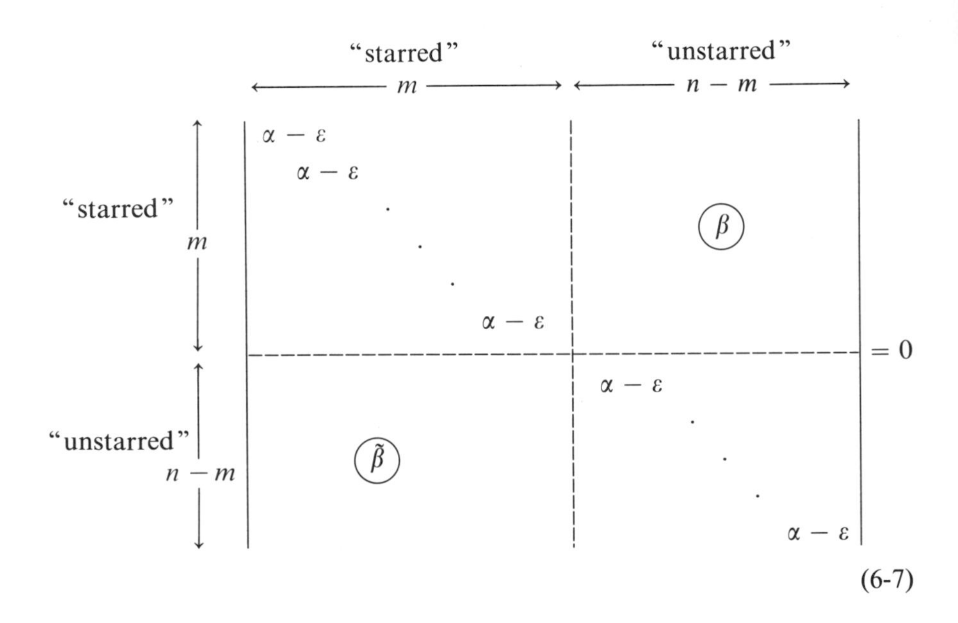

(6-7)

Hence, in the upper-left ($m \times m$) block, there will be the ($\alpha - \varepsilon$)-term along the diagonal but zeros everywhere else. This is so since, by hypothesis, no starred atom is bonded to another starred atom and so all the off-diagonal terms in this ($m \times m$) block (which we might regard as the "starred corner" of the secular determinant) are zero, by assumption b) (equation (6-3)). Any H_{rs}-elements in the first m rows of the secular determinant in (6-7) which are non-zero (*i.e.*, which are β) must occur somewhere within the last ($n - m$) columns of the determinant. This is so because the upper right-hand ($m \times (n - m)$) block represents matrix-elements, H_{rs}, between "starred" and "unstarred" atoms, some of which may be neighbours. We say above, rather vaguely, "somewhere" because we wish to keep the proof general for *any* alternant-hydrocarbon system. Hence we use the symbol (β) in the upper-right ($m \times (n - m)$) block of the determinant to indicate that some elements here will be β and others will be zero. Of course, in anything other than very small molecules there will be more zeros along any one row of this $(m \times (n - m))$ block than β's because, in the sorts of molecules we are considering, a given atom will have only 1, 2, or 3 neighbours. Nevertheless, *some* of the elements in the upper right-hand block will be β, and that is all we need to note for the present. By an exactly similar argument, the lower right-hand $((n - m) \times (n - m))$ block of (6-7) (what we might regard as the "unstarred part" of the secular determinant) will comprise $(\alpha - \varepsilon)$-terms along the diagonal, and zeros everywhere else. Again, the lower left-hand $((n - m) \times m)$ block, $\textcircled{\tilde{\beta}}$, will have some elements β and others zero, since this represents H_{rs} matrix-elements between unstarred and starred atoms. Notice that the lower-left $((n - m) \times m)$ block has been denoted $\textcircled{\tilde{\beta}}$ to indicate that it is the "mirror image", across the main diagonal of the whole determinant, of the upper right $(m \times (n - m))$ block. This must be so because the whole Hamiltonian matrix is (real)-symmetric.

Having obtained the general secular-determinant for an alternant hydrocarbon, we must try to solve it. We want to show that if there is a certain value of ε for which the determinant vanishes, that value being the energy of a molecular orbital, then there is another value of ε, symmetrically related to the first with respect to α, which also makes this determinant zero. This suggests that we should measure energies with α as the reference point and write $y = \varepsilon - \alpha$. The secular determinant then becomes:

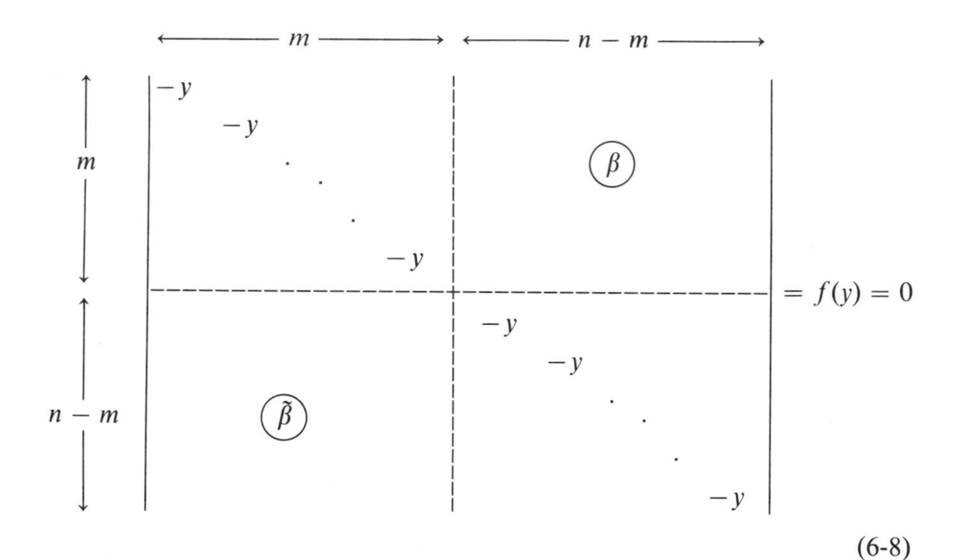

(6-8)

We wish to show that if y_I is a root of the secular determinant (6-8), then $-y_I$ is also a root. If we can do this it will mean that there are two values of ε, symmetrically disposed about α, which satisfy the secular determinant. Let us call the left-hand side of (6-8) $f(y)$—it will be an nth-order polynomial in y, which results when the determinant is developed. We are then interested in the roots of the equation

$$f(y) = 0 \tag{6-9}$$

and require to show that, if (6-9) holds, then

$$f(-y) = 0 \tag{6-10}$$

is also true. In fact, let us consider $f(-y)$; it is obtained by replacing y in (6-8) everywhere by $-y$, *i.e.*,

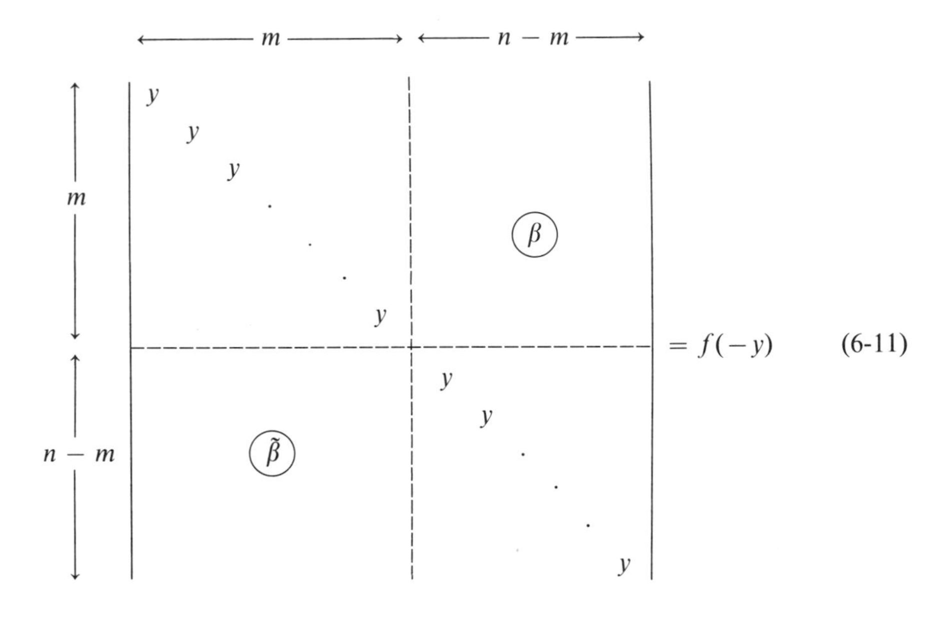

We are now going to do the following:

a) multiply the *top* row (only) by -1 and bring out the factor -1. When this has been done, the first element of the first row is $-y$, those elements amongst the last $(n-m)$ entries of the first row which are non-zero are now $-\beta$, and there is a factor (-1) *outside* the determinant. The remaining $(n-1)$ rows would be entirely unaffected and stand exactly as they are in equation (6-11).

b) We now do the same for *each* of the first m rows. At the end of that process, every element in the first m rows of the determinant will have been multiplied by (-1), and there will be a factor of $(-1)^m$ outside the determinant; all elements of the last $(n-m)$ rows of the determinant will rest unchanged.

Thus:

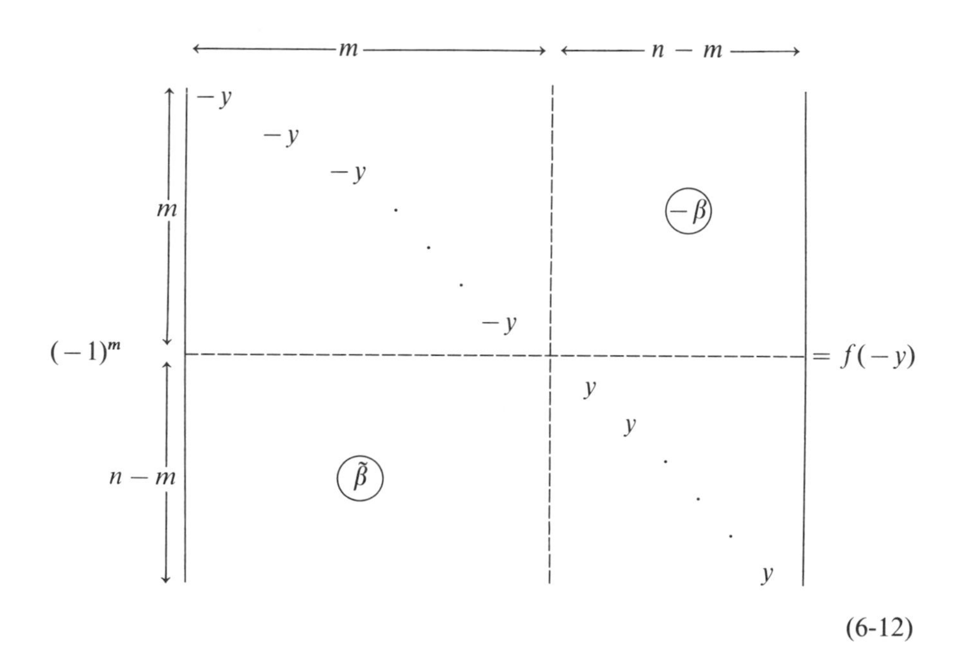

(6-12)

Every β-element in the upper-right $(m \times (n-m))$ block has now therefore been replaced by $-\beta$—this is what the symbol $\bigcirc\!\!\!\!\!-\beta$ is intended to denote.

Now let us repeat the process of multiplication by (-1) but this time do it on certain *columns*, not rows; in fact, we perform this operation on the last $(n-m)$ columns, bringing a factor of (-1) outside the determinant for each column which is so multiplied. After this process has been completed, the upper right-hand $(m \times (n-m))$ block is transformed back to $\bigcirc\!\!\!\!\beta$, the lower right-hand $((n-m) \times (n-m))$ block is a diagonal matrix with $-y$ everywhere along the diagonal, and a further factor of $(-1)^{n-m}$ is brought outside the determinant. Thus, we have, from (6-12):

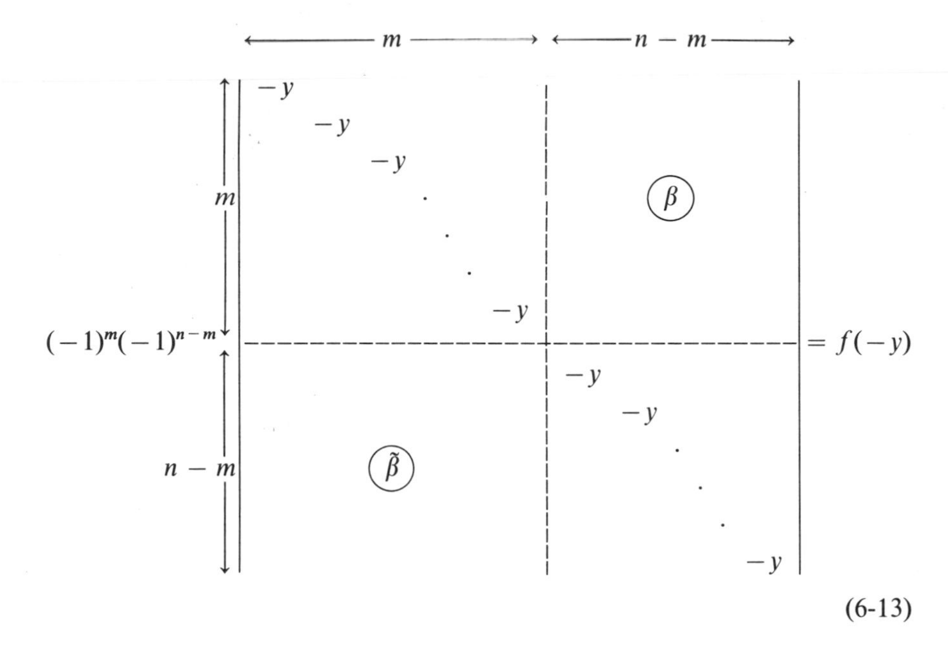

(6-13)

But the determinant in equation (6-13) is just the original determinant we had in (6-8), which was there called $f(y)$. Hence, (6-13) may be written

$$(-1)^n f(y) = f(-y) \tag{6-14}$$

The factor $(-1)^n$ need not concern us for if y_I is a root of the secular determinant (6-8), then $f(y_I) = 0$; hence, if $f(y_I) = 0$, then equation (6-14) tells us that $f(-y_I)$ is also zero and thus that $-y_I$ is also a root of (6-8). Since $y_I = \varepsilon_I - \alpha$ this implies that the two complementary values of ε are $\alpha + y_I$ and $\alpha - y_I$—or, measuring energy in units of β and setting $x_I = y_I/\beta$, $\alpha \pm x_I\beta$. We now see that the energy levels are indeed paired around α as asserted, and hence the Theorem is proved.

Note that the Theorem is true whatever numerical value may be assigned to β. We further see that the Theorem would hold even if all the non-zero H_{rs}-matrix-elements were *not* set at the common value of β, but were assigned different values for different bonds. The arguments outlined between equations (6-6) and (6-14) did not depend on any assumptions about the *magnitudes* of the H_{rs}-terms we were multiplying by (-1) and we now see that restriction c) (equation (6-4)) was not necessary. All that the argument depended on was intelligent and judicious multiplications by (-1)! The Theorem is, therefore, one of very wide generality:

Corollary "Alternant hydrocarbons containing an odd number of carbon atoms must have at least one non-bonding ("zero-energy") molecular-orbital, with energy $\varepsilon = \alpha$".

By an "odd-alternant" hydrocarbon is usually meant a free radical, such as the benzyl radical (Fig. 6-3). In this example there are seven carbon-atoms, hence seven atomic-orbitals and seven molecular-orbitals which must accommodate the seven π-electrons. Of the seven energy-levels, six of them will occur quite naturally in pairs, according to the Theorem we have just proved, as shown schematically in Fig. 6-4. How can the seventh be

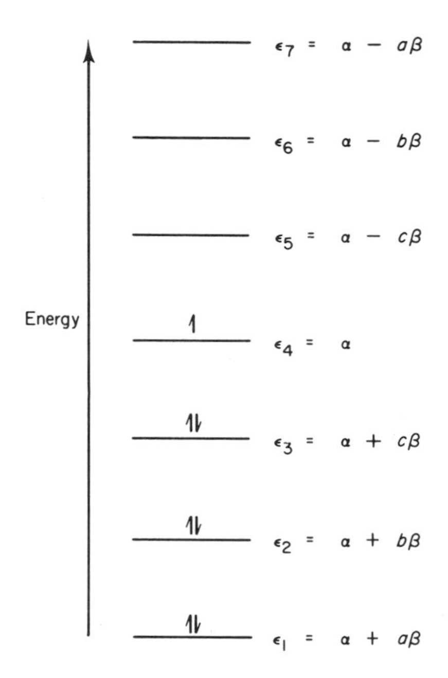

FIG. 6-4 π-Electron molecular-orbital energy-levels and ground-state electronic-configuration of the benzyl radical.

placed so that it has a partner? Only if its partner is the same as itself—*i.e.* if $\alpha + f\beta = \alpha - f\beta$, which requires $f = 0$ and thus $\varepsilon = \alpha$. The "pairing" symmetry of the MO energy-levels is thus preserved by the presence of the "non-bonding" orbital. In this example of the benzene radical, one non-bonding orbital occurred, but in some, rather curious, free-radicals three non-bonding levels arise. This means that two of the other orbitals in the system, which themselves constitute a complementary pair, happen also to be non-bonding orbitals. One could conceive of there being even more than three non-bonding orbitals—perhaps five, for example; in general, however, in an odd-alternant hydrocarbon there must be an *odd* number of such non-bonding orbitals.

6.4 Proof of Part 2 of the Theorem (Concerning the LCAO-Coefficients)

We now prove the second part of the Coulson–Rushbrooke Theorem that in pairs of complementary orbitals the LCAO-coefficients of orbitals centred on the starred atoms are the same, while those of orbitals centred on the unstarred atoms are the same in magnitude but opposite in sign. Let us consider the Ith orbital of energy ε_I (let it be, for argument's sake, a bonding orbital) and its associated set of LCAO-coefficients $\{c_{Ir}\}$, $r = 1, 2, \ldots, n$. These will satisfy a series of secular equations, the rth one of which, on the simple Hückel assumptions, is

$$(\alpha - \varepsilon_I)c_{Ir} + \sum_{\substack{\text{sum over}\\ \text{neighbours}\\ s,\text{ of } r}} \beta c_{Is} = 0 \tag{6-15}$$

Let us start by supposing that r is a starred atom; then all the atoms, s, which are neighbours of r and which appear in the summation of equation (6-15) will be unstarred—if this were not so, the starring process would not have been possible. The argument is again clearer if we measure energies relative to α and in units of β; dividing through (6-15) by β and setting $x_I = (\varepsilon_I - \alpha)/\beta$, we thus obtain

$$-x_I c_{Ir} + \sum_{\substack{\text{sum over}\\ \text{neighbours}\\ s,\text{ of } r}} c_{Is} = 0 \tag{6-16}$$

for the rth secular equation. If we now adopt the convention that MO's are labelled in ascending order of energy, from Ψ_1 (with energy ε_1) denoting the lowest bonding-orbital, to Ψ_n, standing for the highest anti-bonding orbital, the orbital complementary to the Ith bonding-orbital, Ψ_I, will be Ψ_{n-I+1}. For convenience let us call this Ψ_ν with energy ε_ν. Then since ν is the orbital complementary to orbital I,

$$x_\nu = \frac{\varepsilon_\nu - \alpha}{\beta} = -x_I \tag{6-17}$$

Ψ_ν will be characterised by a set of LCAO-coefficients $\{c_{\nu r}\}$, $r = 1, 2, \ldots, n$ which, together with x_ν, will satisfy equations corresponding to (6-16) for Ψ_I, *i.e.*

$$-x_\nu c_{\nu r} + \sum_{\substack{\text{sum over}\\ \text{neighbours}\\ s, \text{ of } r}} c_{\nu s} = 0 \tag{6-18}$$

Since $x_\nu = -x_I$ (equation (6-17)), comparison of equation (6-16) with equation (6-18) shows that the latter would be satisfied if we set

$$\left.\begin{aligned} c_{\nu r} &= c_{Ir} \\ \text{and all} \quad c_{\nu s} &= -c_{Is} \end{aligned}\right\} \tag{6-19}$$

In words, the rth secular equation for the Ith orbital is still satisfied when the energy, ε_ν, of the orbital complementary to I is substituted for ε_I, provided that the coefficient of the atomic orbital centred on the starred atom (r) is kept the same, and the coefficients of those centred on the unstarred atoms are merely changed in sign. What is more, a similar equation will be satisfied for every one of the starred atoms, r, ($r = 1, 2, \ldots, m$).

Now let us go back and assume that r is an *unstarred* atom. Then, by an exactly similar argument, the neighbouring atoms, s, in equations (6-15), (6-16) and (6-18) will be starred vertices. Consider now the same transformation from the Ith orbital to its complement, ν, and again focus attention on the rth secular equations for these two orbitals (equations (6-16) and (6-18) when r is now an *un*starred atom). In equation (6-16), as before, we reverse the sign of x_I, change the sign also of the coefficient of the orbital on the unstarred atom (*i.e.* this time replace c_{Ir} by $-c_{Ir}$) and leave the coefficients of the atomic orbitals centred on the starred atoms (now s) unchanged. This gives

$$(+x_I)(-c_{Ir}) + \sum_{\substack{\text{sum over}\\ \text{neighbours,}\\ s}} c_{Is} = 0 \tag{6-20}$$

which is clearly true since it is identical with equation (6-16). Again, therefore, since $x_\nu = -x_I$ we see that equation (6-18) can be satisfied by setting

$$\left.\begin{aligned} c_{\nu r} &= -c_{Ir} \\ \text{and all} \quad c_{\nu s} &= c_{Is} \end{aligned}\right\} \tag{6-21}$$

A similar argument can be advanced for each one of the $(n - m)$ secular equations for which the running-index r represents the label of an unstarred atom (*i.e.* $r = m + 1, m + 2, \ldots, n$). Thus, taking the quantitites x_I and $\{c_{Ir}\}$, $r = 1, 2, \ldots, n$ (appropriate to the orbital Ψ_I) which satisfy equations like (6-16) for $r = 1, 2, \ldots, n$, and setting

$$\begin{aligned}
&\text{a)}\ x_\nu = -x_I \\
&\text{b)}\ c_{\nu r} = c_{Ir} \text{ if } r \text{ is starred} \\
&\phantom{\text{b)}\ } c_{\nu r} = -c_{Ir} \text{ if } r \text{ is unstarred,} \qquad \text{for all } r = 1, 2, \ldots, n
\end{aligned} \tag{6-22}$$

leads to a set of LCAO-coefficients, $\{c_{\nu r}\}$, $r = 1, 2, \ldots, n$, which satisfy the n secular equations for the complementary-orbital energy, ε_ν. The set of coefficients, $\{c_{\nu r}\}$, $r = 1, 2, \ldots, n$, thus constitutes a molecular orbital corresponding to the energy ε_ν—*i.e.* they are the LCAO-coefficients of Ψ_ν, the orbital complementary to Ψ_I.

Of course, in the above argument, it is immaterial which atoms we originally (and arbitrarily) designated as "starred" and which, as a consequence, are called "unstarred". There is, therefore, nothing magic about having said that it was the coefficients of atomic orbitals centred on the starred atoms which are not altered in the complementary orbital, and the unstarred ones which are—it could equally well have been the other way round. All that we can require from these arguments is that the coefficients of atomic orbitals centred on *one* of the two distinct groups of atoms shall differ in sign between a given orbital and its complement. This is all perfectly consistent with our previous observation that only the *ratios* of the atomic-orbital combinatorial coefficients are obtainable from the variation method and that an LCAO-MO (or, for that matter, any other MO) is determined only to within a constant multiplicative factor (in this case we are talking of a factor of (-1)).

We have thus proved part 2 of the Coulson–Rushbrooke Theorem. Once again, the arguments involved only multiplication by (-1) and so part 2 of the Theorem is also independent of whatever numerical value may be assigned to β. By following the arguments again, starting with the more-general equation (6-6) rather than equation (6-15), the reader may convince himself that part 2 of the Theorem, which we have just proved, also holds if the non-zero matrix elements, H_{rs}, have different values for different bonds, rather than a common value, β, for all bonds as was assumed for simplicity in the above proof. In view of this, assumption c) (equation (6-4)) is unnecessary.

When assumption c) (equation (6-4)) *is* invoked, an amusing result emerges from the preceding analysis. This concerns the LCAO-coefficients of non-bonding orbitals of both alternant and non-alternant systems. It is that the sum of the coefficients associated with all the atoms adjacent to any given atom are zero. This is immediately evident from equation (6-16) for, if the Ith orbital is non-bonding, $x_I = 0$ and hence the sum, $\sum c_{Is}$, over neighbours,

s, of r must be zero. This means, for example, that we can write down straight away the non-bonding MO (Ψ_3) of the (odd, alternant) pentadiene radical (Fig. 6-5). If we start by assigning c_{31} the value f, then, in order that the sum

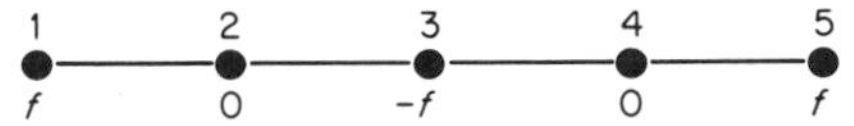

1 2 3 4 5

f 0 $-f$ 0 f

FIG. 6-5 Weighting coefficients of the atomic orbitals centred on carbon atoms 1-5 in the non-bonding MO of the odd, alternant pentadiene-radical.

of coefficients associated with atoms neighbouring atom 1 be zero, c_{32} must be zero. (In this case atom 1 has only one neighbour, *i.e.* atom 2.) For the sum ($c_{31} + c_{33}$) of coefficients associated with atoms (1 and 3) adjacent to atom 2 to be zero, we require c_{33} to have the value $-f$. The non-bonding orbital thus has coefficients

$$f, \quad 0, \quad -f, \quad 0, f$$

or, normalised,

$$\frac{1}{\sqrt{3}}(1, 0, -1, 0, 1).$$

An interesting corollary to part 2 of the Coulson–Rushbrooke Theorem (which does *not* require assumption c), equation (6-4)) concerns the nodes in the various LCAO-MO's discussed in our sample calculation on the alternant hydrocarbon, butadiene (§2.10). As an example, we consider just the highest-bonding and lowest-anti-bonding orbitals of butadiene which, in §2.7, we called Ψ_2 and Ψ_3. These are, of course, complementary orbitals; their nodal behaviour has been redrawn in Fig. 6-6 which is a simplified

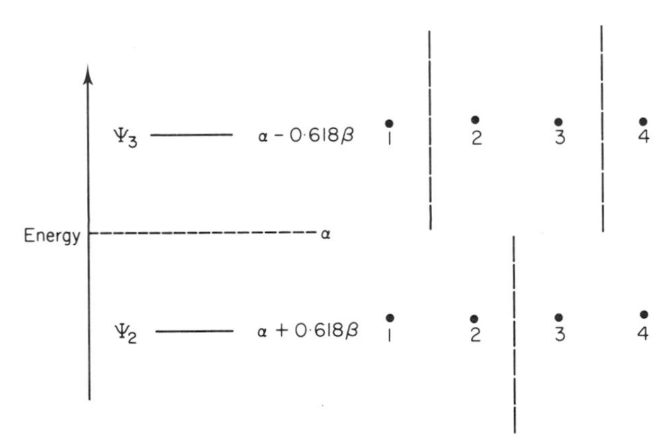

FIG. 6-6 Part of the π-electron, energy-level spectrum of butadiene and the number of nodes in the corresponding molecular-orbitals.

version of Fig. 2-13. Ψ_2 has a single node, between carbon atoms 2 and 3; it is thus bonding in the regions 1-2 and 3-4 but anti-bonding in the 2-3 region. Then, because of the alternation of sign which exists between the LCAO-MO-coefficients of Ψ_2 and Ψ_3 as we proceed along the carbon-atom chain, (according to the Theorem we have just proved), Ψ_3 has two nodes, one in the 1-2 region and one in the 3-4 region; Ψ_3 is thus bonding between carbon-atoms 2 and 3 but anti-bonding in the regions 1-2 and 3-4. There is therefore an *inversion* in bonding and anti-bonding character in the various parts of the molecule which comes into play if an electron is excited from a given orbital to the corresponding complementary orbital. This process (in fact excitation from the highest-occupied to the lowest-unoccupied orbital) is what frequently happens when benzenoid and other condensed systems absorb ultra-violet light. As an example, let us consider the case of naphthalene, (Fig. 6-7).

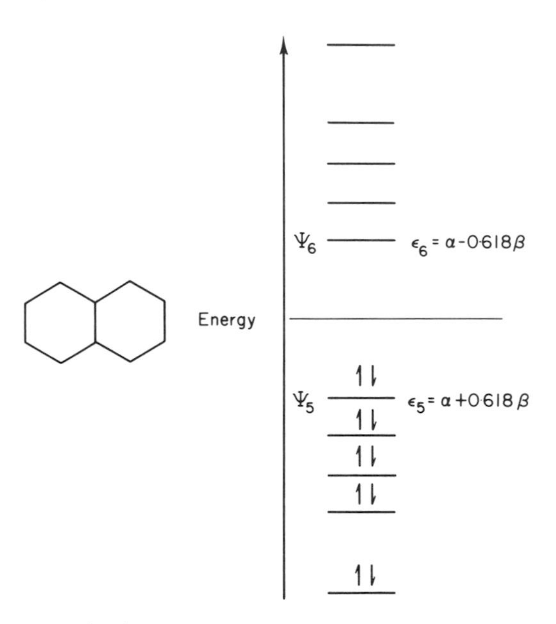

FIG. 6-7 π-Electron, molecular-orbital energy-levels, and ground-state electronic-configuration of naphthalene.

Here, there are ten molecular orbitals which occur in five complementary pairs, as shown schematically in Fig. 6-7. In the ground state, all the five bonding-orbitals are filled, accommodating the 10 electrons in the π-system and leaving all the anti-bonding orbitals empty. The lowest-energy excitation is approximately one from the highest-occupied MO (called Ψ_5 in Fig. 6-7) to the lowest-unoccupied (Ψ_6)[N24]. These are complementary orbitals with energies $\alpha \pm 0{\cdot}618\,\beta$. The long-wave-length absorption in naphthalene thus represents a transition from a bonding orbital to its anti-bonding complement. This will frequently occur in other cases when all the bonding orbitals are filled and all the anti-bonding ones are empty. The example cited is both typical of this situation and, as we shall see shortly (§6.5), interesting because such a transition does not seriously change the distribution of charge within the molecule.

Charge distribution in alternant hydrocarbons is the concern of the third (and main) part of the Coulson–Rushbrooke Theorem, to the proof of which we now turn.

6.5 Proof of Part 3 of the Theorem (Concerning Equality of Atomic Charges)

As a prelude to the proof as such it will be found convenient to cast the Hückel equations in matrix form; as we have continually emphasised, much of Hückel theory is essentially the linear algebra associated with real-symmetric matrices, albeit in a somewhat disguised form. Let us begin by considering again (in the original matrix-notation rather than the "α, β" notation we have lately adopted) the rth secular equation, satisfied by ε_I and $\{c_{Ir}\}$, $r = 1, 2, \ldots, n$. This is

$$(H_{rr} - \varepsilon_I)c_{Ir} + \sum_{s \neq r} H_{rs}c_{Is} = 0 \tag{6-23}$$

This could be written as

$$\sum_{s=1}^{n} H_{rs}c_{Is} = \varepsilon_I c_{Ir} \tag{6-24}$$

where the summation now *includes* the term when $s = r$. Since there are n such equations ($r = 1, 2, \ldots, n$), these could be combined and written as one matrix equation, thus

$$\mathbb{H}\mathbf{C}_I = \varepsilon_I \mathbf{C}_I \tag{6-25}$$

Here $\mathbb{H}$ is a square $n \times n$ matrix $[H_{rs}]$ (equation (2-17)) which represents, in the basis-set with which we are working, the effective, one-electron Hamiltonian, the precise form of which, as we have said, we do not attempt to write down. $\mathbf{C}_I$ is the column vector

$$\mathbf{C}_I = \begin{pmatrix} c_{I1} \\ c_{I2} \\ \vdots \\ c_{In} \end{pmatrix} \tag{6-26}$$

When the equations are written in the matrix form of equation (6-25), it is clear that ε_I is an eigenvalue (or "latent root") of $\mathbb{H}$ and that the set of coefficients, $\mathbf{C}_I$ $(=\{c_{Ir}\}, r = 1, 2, \ldots, n)$, is a corresponding eigenvector (or latent vector) of $\mathbb{H}$. There will be n such eigenvectors, $\{C_I\}$, $I = 1, 2, \ldots, n$, and therefore n (not-necessarily-all-distinct) eigenvalues, ε_I, each pair of which satisfies an equation like (6-25). The n such equations like (6-25) $(I = 1, 2, \ldots, n)$ can now be combined into one, global, matrix-equation by defining two further matrices, $\mathbb{C}$ and $\mathbb{E}$. $\mathbb{C}$ is an $n \times n$ square-matrix whose *columns* are the (normalised) eigenvectors of $\mathbb{H}$ (*i.e.* the n sets of LCAO-coefficients for the conjugated system.)

$$\mathbb{C} = \left(\begin{pmatrix} c_{11} \\ c_{12} \\ \vdots \\ c_{1r} \\ \vdots \\ c_{1n} \end{pmatrix} \begin{pmatrix} c_{21} \\ c_{22} \\ \vdots \\ c_{2r} \\ \vdots \\ c_{2n} \end{pmatrix} \cdots \begin{pmatrix} c_{I1} \\ c_{I2} \\ \vdots \\ c_{Ir} \\ \vdots \\ c_{In} \end{pmatrix} \cdots \begin{pmatrix} c_{n1} \\ c_{n2} \\ \vdots \\ c_{nr} \\ \vdots \\ c_{nn} \end{pmatrix}\right) \tag{6-27}$$

$\mathbb{E}$ is a diagonal matrix in which the various eigenvalues of $\mathbb{H}$, $\varepsilon_1, \varepsilon_2 \ldots \varepsilon_I \ldots \varepsilon_n$, appear along the diagonal, and all off-diagonal elements are zero.

$$\mathbb{E} = \begin{pmatrix} \varepsilon_1 & & & & & & \\ & \varepsilon_2 & & & 0 & & \\ & & \cdot & & & & \\ & & & \cdot & & & \\ & & & & \varepsilon_I & & \\ & 0 & & & & \cdot & \\ & & & & & & \cdot \\ & & & & & & & \varepsilon_n \end{pmatrix} \tag{6-28}$$

(Notice that the order in which the eigenvalues of $\mathbb{H}$, $\{\varepsilon_I\}$, $I = 1, 2, \ldots, n$, appear along the diagonal in $\mathbb{E}$ *must* be the order in which the corresponding eigenvectors[N29] of $\mathbb{H}$, $\{\mathbf{C}_I\}$, $I = 1, 2, \ldots, n$, appear as columns (from left to right) in $\mathbb{C}$.)

The global matrix-equation which takes in the n equations like (6-25), $(I = 1, 2, \ldots, n)$ is then

$$\mathbb{H}\mathbb{C} = \mathbb{E}\mathbb{C} \tag{6-29}$$

For our proof of part 3 of the Coulson–Rushbrooke Theorem we now focus attention on the matrix $\mathbb{C}$ (equation (6-27)). We are assuming that the LCAO-MO coefficients are a) normalised, *i.e.*

$$\mathbf{C}_I \cdot \mathbf{C}_I = \sum_{r=1}^{n} c_{Ir} c_{Ir} = 1 \tag{6-30}$$

(This was our normalisation condition of equation (2-34)) (b) orthogonal, *i.e.*,

$$\mathbf{C}_I \cdot \mathbf{C}_J = \mathbf{C}_J \cdot \mathbf{C}_I = \sum_{r=1}^{n} c_{Ir} c_{Jr} \quad (I \neq J) = 0 \tag{6.31}$$

(This condition will be *automatically* true if $\{\varepsilon_I\}$, $I = 1, 2, \ldots, n$, contains no degeneracies and can always be *arranged* to be true even if degeneracies in $\{\varepsilon_I\}$ do occur—see note 25). These two conditions can be unified into one algebraic equation

$$\mathbf{C}_I \cdot \mathbf{C}_J = \sum_{r=1}^{n} c_{Ir} c_{Jr} = \delta_{IJ} \tag{6-32}$$

or one matrix-equation

$$\tilde{\mathbb{C}}\mathbb{C} = \mathbb{1}_{n \times n} \tag{6-33}$$

where $\tilde{\mathbb{C}}$ is the transpose of $\mathbb{C}$ and $\mathbb{1}_{n \times n}$ is the unit matrix of order n. Since, by definition,

$$\mathbb{C}^{-1}\mathbb{C} = \mathbb{1}_{n \times n} = \mathbb{C}\mathbb{C}^{-1} \tag{6-34}$$

where $\mathbb{C}^{-1}$ is the inverse of $\mathbb{C}$, it follows from (6-33) that

$$\tilde{\mathbb{C}} = \mathbb{C}^{-1} \tag{6-35}$$

and hence, from the right-hand side of (6-34), that

$$\mathbb{C}\tilde{\mathbb{C}} = \mathbb{1}_{n \times n} \tag{6-36}$$

By equating diagonal elements of the matrices on either side of equation (6-36), we obtain

$$\sum_{I=1}^{n} c_{Ir}^2 = 1 \tag{6-37}$$

This is the crucial result which will enable us, within a few lines, to prove part 3 of the Coulson–Rushbrooke Theorem. Those readers who have preferred not to follow in detail equations (6-23)–(6-37) may, by accepting the validity of the latter equation, pick up the argument again from here.

Let us give some consideration to what equation (6-37) means physically. It means, according to equation (4-4), that if we were in some way able to put *just one* electron into *every* MO—both bonding and anti-bonding—of a conjugated hydrocarbon-system, then the charge on each carbon-atom would be precisely unity; of course, chemically, this is a ridiculous thing to suggest—however, the statement will be useful to us, as will be seen presently. Note that we have not said anything yet about the given conjugated system's being alternant or non-alternant; therefore, equation (6-37) and the statement in the previous sentence are *true for both classes of hydrocarbons.*

Let us now specialise the argument to the case of an even, alternant hydrocarbon with $n/2$ doubly-occupied bonding-orbitals and $n/2$ unoccupied anti-bonding orbitals. This is the situation, for example, in the ground state of butadiene (Fig. 2-8) and of naphthalene (Fig. 6-7). From equation (4-4), the charge on the rth carbon atom, q_r, is

$$q_r = 2\sum_{I=1}^{n/2} c_{Ir}^2 \tag{6-38}$$

Were r to be a starred atom, we know from part 2 of the Coulson–Rushbrooke Theorem (§6-4) that

$$c_{Ir} = c_{(n-I+1),r} \tag{6-39}$$

Equation (6-38) may therefore be partitioned into two terms, as follows:

$$q_r = \sum_{I=1}^{n/2} c_{Ir}^2 + \sum_{I=1}^{n/2} c_{Ir}^2 = \sum_{I=1}^{n/2} c_{Ir}^2 + \sum_{n/2+1}^{n} c_{Ir}^2 \tag{6-40}$$

These two summations can then be recombined and equation (6-37) called upon, to give

$$q_r = \sum_{I=1}^{n} c_{Ir}^2 = 1 \tag{6-41}$$

Were r an unstarred atom, then equation (6-39) is replaced by

$$c_{Ir} = -c_{(n-I+1),r} \tag{6-42}$$

and equation (6-40) then becomes

$$q_r = \sum_{I=1}^{n/2} c_{Ir}^2 + \sum_{I=n/2+1}^{n} (-c_{Ir})^2 \tag{6-43}$$

which again leads to (6-41).

Hence, because of the result expressed in equation (6-37), the pairing of MO's and the consequent symmetry of LCAO-coefficients between pairs of complementary orbitals, the charge density on the rth carbon-atom of a neutral, even, alternant hydrocarbon in its ground state, is unity. This is the essence of the third part of the Coulson–Rushbrooke Theorem. Its proof depends on the fact that the square of a quantity is the same as the square of minus that quantity, and is thus seen to be a natural consequence of parts 1 and 2 of the Theorem.

The Theorem also applies to neutral radicals which are odd, alternant hydrocarbons. Let us take, as an example, the benzyl radical (Figs. 6-3 and 6-4) with seven LCAO-MO energy-levels, in three complementary bonding- and anti-bonding pairs and one, lone, non-bonding orbital. In the ground state, the three bonding-orbitals will be doubly occupied, the non-bonding orbital will be singly occupied, and the anti-bonding orbitals will be empty (Fig. 6-4). If, as is customary, the orbitals are labelled in order of increasing energy, the bonding orbitals will be $\Psi_1, \Psi_2, \ldots, \Psi_{(n-1)/2}$, the anti-bonding orbitals will be $\Psi_{(n+3)/2}$ to Ψ_n and the non-bonding orbital will be $\Psi_{(n+1)/2}$. Under these circumstances, equation (4-4) determines that the charge on the rth carbon-atom should be

$$q_r = 2\left(\sum_{I=1}^{(n-1)/2} c_{Ir}^2\right) + c_{(n+1)/2,r}^2 \tag{6-44}$$

By arguments entirely analogous to those outlined between equations (6-38) and (6-43), equation (6-44) can be written

$$q_r = \left(\sum_{I=1}^{(n-1)/2} c_{Ir}^2\right) + \left(\sum_{I=(n+3)/2}^{n} c_{Ir}^2\right) + c_{(n+1)/2,r}^2$$
$$= \sum_{I=1}^{n} c_{Ir}^2 = 1, \tag{6-45}$$

the last step again by equation (6-37).

Hence, the Theorem is proved also for neutral, alternant, radicals. Of course, it does not apply to radical cations and anions in which there is a net charge-distribution, as may be verified experimentally from the reactions which they undergo.

It is also evident that the Theorem will not apply to non-alternant hydrocarbons because the basis on which we were able to separate the charge distribution into two halves, one to the bonding- and the other to the anti-bonding orbitals, will no longer be valid. This means that non-alternant systems, even though they may consist only of carbon and hydrogen, must be expected to display charge migration and thus non-zero dipole-moments. In one sense, therefore, a Hückel calculation on a neutral, alternant hydrocarbon may be said to be "self-consistent" in that each atomic orbital in the LCAO-scheme contributes one electron to the π-system and, at the end of the day when the LCAO-MO's have been formed, any given carbon-atom (on which an atomic orbital participating in the linear combination is centred) has precisely unit electronic charge on it. In other words, the charge density we calculate is consistent with the assumptions on which the calculation was based. A Hückel calculation on a non-alternant hydrocarbon, on the other hand, is not "self-consistent" in this sense; some carbon atoms have more, and others have less, electronic charge than was originally contributed by the atomic orbitals centred on them. As an example, Fig. 6-8 shows the calculated charge-distribution on the various non-equivalent carbon-atoms in naphthalene and its non-alternant isomer, azulene; naphthalene and azulene are, historically, one of the test cases for discussions of this sort.

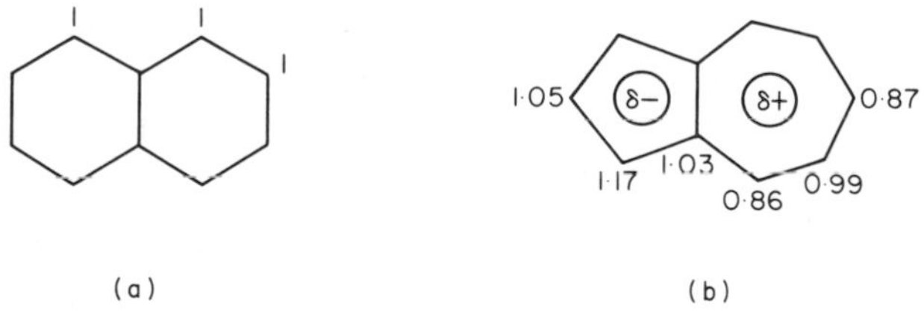

FIG. 6-8 Molecular diagrams of (a) naphthalene and (b), its non-alternant isomer, azulene.

Figure 6-8b thus illustrates, in a numerical rather than a qualitative way, the argument about the $4p + 2$ rule outlined in §5.3. We said there that the five-membered ring would tend to collect charge whilst the seven-membered ring would be more stable if it gave charge away; hence, in a somewhat ill-defined way, we were able to use symbols like $\delta-$ and $\delta+$ (Figs. 5-6 and 5-7) in order to show that the five-membered ring of azulene was, on the whole, negatively charged and the seven-membered ring positively charged. The numerical values for charges indicated in Fig. 6-8 fit this qualitative picture perfectly well; every atom in the five-membered ring bears more than unit electronic charge and, equally, the seven-membered ring has lost charge because the charges on the carbon atoms in that ring sum to less than the equivalent of seven π-electrons. To this extent, therefore, we have now quantitatively substantiated the arguments put forward in a somewhat *ad-hoc* manner in §5.3.

Thus far we have shown that this uniform charge-distribution applies to the ground states of neutral, alternant hydrocarbons and neutral, alternant radicals. If we consider excited states of these species then the Theorem concerning charges is no longer necessarily true; the Theorem is, however, still true for excited states *if the excitation takes place from one orbital to its complementary partner*; this will, in fact, normally be the longest-wave-length absorption and is typified, for example, by the naphthalene transition discussed at the end of §6.4. Since, by part 2 of the Coulson–Rushbrooke Theorem (§6.4), the coefficients of an antibonding orbital are obtained from those of its complementary bonding-orbital merely by changing the *signs* of the coefficients associated with either the starred or the unstarred set of atoms, and since the charge on a given atom is obtained from the *square* of such LCAO-coefficients (equation (4-4)), it is immediately evident that the charge distribution associated with an (occupied) anti-bonding orbital is the same as that contributed by its (similarly occupied) complementary bonding-orbital. By examining the LCAO-coefficients for butadiene in equation (2-67) (§2.7), the reader may easily verify that the configuration $(\Psi_1)^2(\Psi_2)^1(\Psi_3)^1$ of butadiene, in which an electron has been excited from Ψ_2 to its complementary orbital, Ψ_3, has the same charge distribution as the ground state,

$(\Psi_1)^2(\Psi_2)^2$, but that an excitation such as, for example, $\Psi_2 \rightarrow \Psi_4$, giving the configuration $(\Psi_1)^2(\Psi_2)^1(\Psi_4)^1$, would cause charge to be shifted around the molecule during the absorption; Ψ_2 and Ψ_4 are not complementary orbitals.

Finally, we note that, once more, we have not made any specific reference to β in proving part 3 of the Coulson–Rushbrooke Theorem; hence, this part of the Theorem also does not depend on the numerical value assumed for β, nor does it require that the resonance integrals, β_{rs}, should be the same for all bonds—*i.e.*, assumption c) (equation (6-4)) is again not necessary.

6.6 Assumptions on Which All Three Parts of the Coulson–Rushbrooke Theorem Depend

We see therefore that no part of the Coulson–Rushbrooke Theorem on alternant hydrocarbons depends on having all non-zero Hamiltonian matrix-elements, H_{rs}, equal. In order for the reader to be quite clear which assumptions, in the context of the simple Hückel-method, *are* necessary for the Theorem to hold, we summarise them again below. We require

1) All Coulomb integrals to be the same *i.e.*,

$$\alpha_r = \alpha, \qquad r = 1, 2, \ldots, n \tag{6-46}$$

2) All non-neighbouring Hamiltonian matrix-elements to be zero *i.e.*,

$$H_{rs} = 0, \qquad r \neq s \quad r, s \textit{ not } \text{bonded} \tag{6-47}$$

3) Overlap between different atoms to be neglected *i.e.*,

$$S_{rs} = \delta_{rs} \tag{6-48}$$

Although the entire Theorem is thus founded on all but one of the simplest Hückel-approximations which we have been discussing so far, parts 1 and 3 of the Theorem (concerning pairing of the energy levels, and charges in the ground state, respectively) do carry through into the more-elaborate π-electron theory of Pariser, Parr and Pople,[R17] briefly referred to in Chapter Nine.

6.7 A Mathematical Note

Finally, the reader was advised at the beginning of the present chapter that it would be more formal and mathematical than most of the other chapters in this book; appropriately, therefore, we close on a mathematical note by drawing attention to the fact that parts 1 and 2 of the Coulson–Rushbrooke Theorem are really nothing more than a special case of the famous Perron–Frobenius Theorem on non-negative matrices.

A more abstract, matrix-proof which, in essence, derives parts 1 and 2 of the Coulson–Rushbrooke Theorem in just one operation, is therefore presented, for the interested reader, in Appendix D.

Seven

Improvements and Refinements to the Simple Hückel Method

7.1 Introduction

By means of the straightforward Hückel-method described in Chapter Two, we saw, in §6.5, that a five-membered ring tends to collect charge, whilst a seven-membered ring loses it. However, the numerical values of the charges calculated by the simple Hückel-method are too large; we can tell that they are too large because we can use them to estimate dipole moments. Let us illustrate this by means of a particularly simple example—that of a diatomic molecule with charges $+q$ and $-q$ separated by a distance R (Fig. 7-1). The dipole moment of this system, μ, is given by

$$\mu = qR \tag{7-1}$$

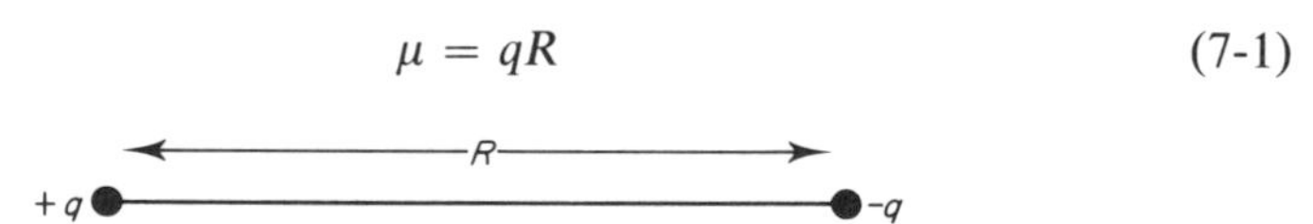

FIG. 7-1 Diatomic molecule with charges $+q$ and $-q$ separated by a distance, R.

If the π-electron charges for azulene given in §6.5 are used, in this kind of formulation, to compute the π-electron dipole-moment of azulene, it is found that the calculated moment is three or four times larger than the observed one. This kind of observation leads one to start asking questions as to whether there are ways in which the simple Hückel-theory can be altered or modified so that it is able to cope, at least in part, with some of these difficulties. These are the questions to which we address ourselves in the succeeding sections of this chapter. The discussion involves primarily modifications to the Coulomb integrals (§§7.2 and 7.3) but we shall also give some attention (§7.4) to how initial estimates of resonance integrals may be improved, and to simultaneous variation of these two types of integrals (§7.5).

7.2 Modification of the Coulomb Integrals: The ω-Technique

The simplest of all these improved Hückel-theories is called the ω-technique. This goes back (despite what is often said to the contrary) to Wheland and Mann who used it, long ago[R18], to discuss the molecule which we have just referred to—azulene. It was then developed, somewhat independently, by the late W. E. Moffitt in his D.Phil. Thesis submitted to the University of Oxford[R19] and subsequently by Streitwieser[R20], with whose name it is now more frequently associated.

We should begin by asking about one of the mistakes, in particular, which Hückel theory would make when, in a hydrocarbon, for example, the charges are not all uniform; as we have seen, in §6.5, this situation would obtain in a non-alternant hydrocarbon. In such a case, the particular error in question is associated with the Coulomb term for atom r. The reader will recall that this was defined in (§2.5) to be the integral

$$\alpha_r = \int \phi_r \mathscr{H}_{\text{effective}} \phi_r \, d\tau, \tag{7-2}$$

$\mathscr{H}_{\text{effective}}$ being a somewhat-nebulous one-electron Hamiltonian, the precise form of which we do not attempt to write down (see §2.2). We could then argue that if a given atom, r, of the conjugated system had an excess of electrons associated with it (*i.e.* if, in a hydrocarbon for example, q_r were greater than unity) then any other electron which came near this atom r would experience particularly severe additional repulsions because of the familiar Coulomb interactions which like charges exert on one another. Crudely, therefore, a superfluity of electrons in the vicinity of a given atom makes the approach of further electrons that much more difficult. Alternatively, reversing the above argument, we could say that, in this situation, it is easier to remove an electron which *is* already there than it would be if atom r were positively charged (*i.e.* if q_r were less than unity). This kind of argument can be verified experimentally for it is possible to measure the ionisation potential of an atom—for example, of an oxygen atom in a carbonyl group (Fig. 7-2); in this situation, the ionisation potential of a so-called "non-bonding" electron on the oxygen atom can be studied. Now if there are electron donors (for example, methyl groups) in other parts of the

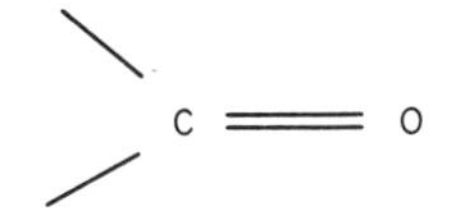

FIG. 7-2 Carbonyl group.

molecule then these will tend to push electrons towards the oxygen atom; experiment then reveals that the ionisation potential diminishes. On the other hand, when the $—CH_3$'s are replaced by less-electron-donating groups, the ionisation potential increases—and if electron-withdrawing groups are present it will increase even more. This therefore constitutes experimental evidence to show that the binding of an electron on an atom depends not only on what atom it is, but also on how many electrons have been pushed onto it, or taken away from it, by the rest of the molecule.

Returning, therefore, to consideration of a carbon atom (r) in a non-alternant hydrocarbon, we see that if, on atom r, there is a piling-up of electrons, we should expect that electron-removal would be a comparatively easy process and that, consequently, the quantity α_r would be numerically smaller than the standard carbon-value, α. This is so because (as was argued in §2.2) α_r is approximately equal to the ionisation potential of an electron on atom r—that is to say, the energy required to take an electron from atom r to infinity.

The proposal which Wheland and Mann first made, therefore, was that the α_r should not be set as a constant for each particular type of atom, irrespective of its position or its rôle in the conjugated system, but it should be allowed to vary, from atom to atom, according to the π-electron charge which each atom carries. The form of the functional dependence of α_r on q_r which is more obviously and immediately sensible is a linear one, although, if one wished to be more refined, a more elaborate expression could be devised. We therefore write:

$$\alpha_r(q_r) = \alpha_r(1) - \omega\beta(q_r - 1) \tag{7-3}$$

α_r thus depends on (*i.e.* is a function of) q_r, by which is meant the *number* of electrons on atom r—for let us be clear that q_r represents a *number* of π-electrons (though not necessarily—in fact, not usually—integral, of course). $\alpha_r(1)$ thus represents the Coulomb integral for a *neutral* atom (r) of *this type* (carbon, nitrogen, etc.), that is, one with charge $q_r = 1$. The term $\omega\beta(q_r - 1)$ may then be regarded as a "correction" to $\alpha_r(1)$ which depends upon the extent to which the charge on atom r, q_r, is different from unity. The proportionality factor, ω, is conveniently given in units of β. Inclusion of the factor β is therefore necessary in order to bring the expression on the right-hand-side of (7-3) back to the dimensions of energy. Further, since β is negative, and ω is conventionally positive, this requires the factor $\omega\beta(q_r - 1)$ to appear in (7-3) preceded by a *negative* sign. A value of *ca.* 1.4 is usually taken for the constant proportionality-factor, ω. There is no very strong theoretical basis for this, merely an empirical one founded on a study of ionisation potentials; indeed, for some purposes, a smaller value would actually be more appropriate.

The consequence of all this discussion is that when we now come to set up the secular determinant (equation (2-50)) for any arbitrary, conjugated, hydrocarbon-system, we cannot simply write "$\alpha_r - \varepsilon$" as the diagonal element in the rth row and rth column without saying that this depends upon the total number of π-electrons on the rth atom. In a sense, we are therefore now confronted with a circular argument; for we cannot even write down the value of α_r until we know what q_r is and yet we will not know what q_r is until we have inserted a value of α_r into the secular equations, solved them and then determined the LCAO-MO coefficients and thence the π-electron atomic-charges, $\{q_r\}$, $r = 1, 2, \ldots, n$. Circular situations of this kind are always dealt with by some kind of *iterative* process. In the present case, the procedure for a neutral hydrocarbon is as follows; first of all, a simple, unrefined Hückel-calculation is performed—in other words, a set of secular equations in which all α_r are set equal to the standard carbon-value, α, is solved to give a set of simple Hückel-charges, $\{q_r\}$, $r = 1, 2, \ldots, n$. On the basis of this simple HMO-charge for atom r, a first-refined estimate of α_r (let us call it $\alpha_r^{(1)}$) can then be made *via* equation (7-3); this can be done for each atom r, $r = 1, 2, \ldots, n$ and hence a modified set of secular equations can be written down with $(\alpha_r^{(1)} - \varepsilon)$ terms along the diagonal. When *these* equations have been solved, a second estimate of the charges, $\{q_r^{(1)}\}$, $r = 1, 2, \ldots, n$, can be made on the basis of the new LCAO-coefficients. Then, again from equation (7-3), (and extending, in a self-evident way, the above notation) we can set

$$\alpha_r^{(2)} = \alpha_r(1) - \omega\beta(q_r^{(1)} - 1), \qquad r = 1, 2, \ldots, n \tag{7-4}$$

and repeat the secular-equation solution yet again to yield $\{q_r^{(2)}\}$, $r = 1, 2, \ldots, n$. The whole cyclic process is then continued until

$$\left.\begin{aligned} q_r^{(m)} &= q_r^{(m-1)} \\ \text{and} \qquad & \\ \alpha_r^{(m)} &= \alpha_r^{(m-1)} \end{aligned}\right\} \text{for all } r, \qquad r = 1, 2, \ldots, n \tag{7-5}$$

to within a specified degree of accuracy. When this situation arises (*i.e.* when the iteration has *converged*) the $\{\alpha_r\}$ in the secular determinant have given rise to $\{q_r\}$ which in turn are consistent, *via* the recipe detailed in equation (7-3), with the $\{\alpha_r\}$ upon which their calculation was based. This kind of iterative process is particularly amenable to solution by means of extensive computer programs which can handle such problems routinely. So, in practice as well as in principle, there is no difficulty in executing a calculation scheme of this sort.

Now let us examine the effect which this procedure has on the actual numerical values of the charges in a specific case; we look again at the azulene molecule (Fig. 6-8).

TABLE 7-1

Charges in Azulene

	q_r					
r	1	2	3	4	5	6
Simple Hückel	1·047	1·173	1·027	0·855	0·986	0·870
Iterative ("ω") Hückel	0·955	1·116	1·020	0·907	0·977	0·918

In the first row of Table 7-1 are listed the charges on the six non-equivalent carbon-atoms of azulene, calculated by the simple, unrefined Hückel-method, as described in Chapter Two and §6.5. The second row of the Table contains the corresponding quantities obtained after carrying through the "ω-technique" iteration to completion. Notice that the charges are now much more uniform; in the simple HMO-calculation the maximum charge-difference between two carbon-atoms is $(1{\cdot}173 - 0{\cdot}855) \approx 0{\cdot}32$, while the corresponding quantity from the results of the ω-technique calculation is $(1{\cdot}116 - 0{\cdot}907) \approx 0{\cdot}21$.

This process alone is not, of course, sufficient to overcome all the semi-quantitative shortcomings of the simple HMO-method and, indeed, we would not expect it to be. It does, however, reduce the dipole moment to a much more manageable value, although this quantity is still overestimated by the ω-technique calculation; we should certainly therefore need to look at some further improvements (§§7.3 and 7.4) before we were anything like satisfied with this sort of approach. The ω-technique does, however, give rise to a more realistic predicted charge-distribution than does the simple Hückel-method *in the case of those molecules in which all the π-charges are not unity.*

This latter statement leads us to a very simple but important point (hinted at in §6.5) on which the reader may care to ponder whilst ending the present section. That is that, if all the charges, q_r, in a given molecule were unity, then the "correction" term, $\omega\beta(q_r - 1)$, in equation (7-3) would be precisely zero. As we saw from the Coulson–Rushbrooke Theorem in §6.5, all charges *are* unity in alternant hydrocarbons (such as naphthalene, anthracene and so on); hence, it is when applied to these situations—*i.e.* neutral hydrocarbons in their ground states—that the simple Hückel-method is on its most secure footing. To these molecules, therefore, the objections raised at the beginning of this section do not apply and the ω-technique is irrelevant in such cases, for there is then a sense in which we can say that the simple HMO-theory is "self-consistent" with respect to the Coulomb integrals. However, as will be appreciated from the discussion given in §6.5, the objections outlined above may still hold for certain excited states (those in which a transition has taken place between *non*-complementary orbitals) even if the basic molecule under consideration *is* alternant.

7.3 Further Refinements to the ω-Technique: Kuhn's Method

In 1963, Kuhn[R21] proposed an extension to the simple ω-technique approach just described which enables account to be taken not only of the effect on a given Coulomb-integral of the charge on the corresponding atom, but of the effect of charges on first- and second-neighbours of the atom in question. Thus, Kuhn proposed setting

$$\alpha_r = \alpha_r(1) - \omega\beta(q_r - 1) - \omega'\beta \sum_{\substack{\text{all first-}\\ \text{neighbours,}\\ f,\text{ of } r}} (q_f - 1) - \omega''\beta \sum_{\substack{\text{all second-}\\ \text{neighbours,}\\ s,\text{ of } r}} (q_s - 1) \tag{7-6}$$

Building on the ideas of Wheland and Mann, Kuhn supposed that the (constant) parameters ω, ω' and ω'' represented the electrostatic potentials created at vertex (atom) r of the conjugated system by the charge localised on atoms r, f and s, respectively. Kuhn adopted the conventional value for ω of 1.4 and then considered ω' and ω'' to be proportional to ω and inversely proportional to the distance between the pairs of centres, r and f, and r and s, respectively. Thus[R21, R22]

$$\left.\begin{aligned} \omega &= 1{\cdot}4 \\ \omega &= 1{\cdot}4/1{\cdot}5 = 0{\cdot}9333 \\ \omega &= 1{\cdot}4/2{\cdot}2 = 0{\cdot}6364 \end{aligned}\right\} \tag{7-7}$$

Once equation (7-6) has been established, the iterative scheme then proceeds exactly as in the simple ω-technique approach described in the previous section, except that the dependence of $\{\alpha_r\}$, $r = 1, 2, \ldots, n$, on $\{q_r\}$, $r = 1, 2, \ldots, n$, is now represented by n equations like (7-6) rather than by n equations of the form (7-3).

Opinions on the utility of this extra refinement seem to be diverse; Gayoso has argued[R22] strongly for Kuhn's method, while Coulson and Wille[R23] are sceptical about whether the extra labour involved in the calculations justifies any improvements in the results; indeed, they concluded that, in many cases, this ω-ω'-ω'' technique does not cause any worthwhile improvement in the numerical values of LCAO-MO coefficients and energy-levels obtained from the calculations.

7.4 Modification of the Resonance Integrals: The Coulson–Gołebiewski Approach

We saw at the end of Chapter Two (§2.11) that the resonance integral, β_{rs}, for a given bond between atoms r and s, must be directly related to the *length* of that bond, d_{rs}. As long ago as 1937, Lennard–Jones recognised[R24] that β_{rs} could be expressed as a parabolic function of d_{rs}, while, more recently, Longuet-Higgins and Salem[R25] have proposed that β_{rs} should be represented as an inverse-exponential function of d_{rs}. Now from the discussion of §4.3(c) we know that d_{rs} is explicitly a function of p_{rs}, the bond order of the bond $r - s$. By means of, for example, an expression of the form

$$\beta_{rs} = \beta\, e^{(ap_{rs}+b)} \tag{7-8}$$

it is thus possible to express resonance integrals entirely in terms of the corresponding bond-orders, without having to make appeal to physically observable quantities such as bond length; such quantities are, strictly, outside the framework of the theory *per se*, in the sense that they are not directly accessible *via* quantities—LCAO-MO coefficients and energy-levels—obtained from the secular determinant. This form of expression was adopted by Coulson and Gołebiewski[R26] who found empirically that values of 0·48294 and 0·32196 for the constants a and b, respectively, were suitable for use in iterative processes analogous to those employed in the ω-type techniques (§§7.2 and 7.3).

We observed in the previous section that, in the case of neutral, alternant hydrocarbons in their ground states, since the charge densities on all carbon-atoms are uniform, iterative correction of the Coulomb integrals by any form of ω-technique is irrelevant. However, in any hydrocarbon in which not all the carbon-carbon bonds are equivalent (and this of course means all hydrocarbons except annular ones in the simple HMO-treatment—see §2.11) some form of "self-consistency" is appropriate between the non-zero off-diagonal elements (*i.e.* the resonance integrals, β_{rs}) of the Hückel Hamiltonian and the corresponding bond-orders, p_{rs}, calculated from that Hamiltonian. This can be achieved in an iterative process entirely analogous to that described in §7.2. In the first cycle, all Coulomb-integrals are set at the standard value of α, and all resonance-integrals are given the common value, β. In succeeding cycles (l) of the iterative scheme, Coulomb integrals remain at α (if, as is assumed here, an alternant hydrocarbon is being dealt with), and the resonance integrals, $\beta_{rs}^{(l)}$, are re-estimated from equation (7-8) on the basis of the bond orders, $p_{rs}^{(l-1)}$, calculated in the previous cycle. The process is continued until

$$\left.\begin{aligned} p_{rs}^{(m)} &= p_{rs}^{(m-1)} \\ \text{and} \qquad & \\ \beta_{rs}^{(m)} &= \beta_{rs}^{(m-1)} \end{aligned}\right\} \text{ for all bonds, } (r - s) \qquad (7\text{-}9)$$

to within a predetermined and specified criterion. When the iteration has thus been deemed to have converged, the current set of $\{\beta_{rs}\}$ which constitute the non-zero off-diagonal elements of the Hückel Hamiltonian-matrix give rise to a corresponding set of bond orders, $\{p_{rs}\}$, which are identical to those used to estimate the $\{\beta_{rs}\}$; hence the desired "self-consistency" (in this sense) between resonance integrals and bond orders has been achieved.

7.5 Simultaneous Variation of Coulomb and Resonance Integrals

The obvious and logical extension to the discussion so far outlined in this chapter is to ask whether, in the case of, say, a general, non-alternant hydrocarbon, it is possible and legitimate to perform an iterative calculation in which both Coulomb integrals (§§7.2 and 7.3) *and* resonance integrals (§7.4) are varied simultaneously. In such a scheme, calculations based on relations (7-6) and (7-8) would be carried out in an iterative fashion such that the one-electron Hamiltonian-matrix $[\alpha_r, \beta_{rs}]$ furnished $\{q_r\}$ and $\{p_{rs}\}$ identical with those which had served to calculate the particular $\{\alpha_r\}$- and $\{\beta_{rs}\}$-elements in question; such LCAO-MO coefficients and energy-levels as were derived from this process would then in principle be truly "self-consistent", in the sense implied. This approach has been called[R22] the "self-consistent Hückel-method" or the "$\beta\omega'\omega''$ method".

It is worth noting, in conclusion, that iterative procedures which modify both types of integral simultaneously have been criticised[R27] on the grounds that they could lead to an incorrect convergence limit. Such questions are really outside the scope of the present discussion and we merely refer the reader to an exhaustive investigation of them by Gayoso[R22]. Certainly, these "simultaneous" iterative-procedures are often very slow to converge to the limit which they *do* finally settle on and methods have been suggested for accelerating the convergence[R23] in such cases. The rate-determining factor in the convergence is usually the self-consistency of the charges and Coulomb integrals (which often requires 30 or more iterations[R16]), rather than that of the resonance integrals and bond orders, which frequently reach an equilibrium after only about ten cycles.

Eight

Molecular Diagrams and Reactivity

8.1 Molecular Diagrams

In Chapter Four and elsewhere we have seen how to obtain a considerable amount of information about a molecule—its charges, bond orders and free valences. It is convenient to gather these together into what has come to be called a *molecular diagram*. This is merely a conventional way of representing these data in a simple, visual form. Let us take the case of aniline (Fig. 8-1).

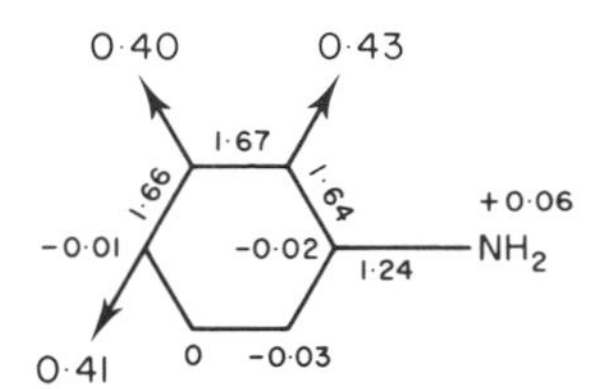

FIG. 8-1 Molecular diagram for aniline.

The bond orders (1·24, 1·64, 1·67 and 1·66) are written, naturally enough, along the bonds; only the values for non-equivalent bonds are explicitly included in the diagram—the correspondingly related ones will be equal by symmetry. Similarly, the charges (+0·06, −0·02, −0·03 and −0·01) are written at the various non-quivalent atoms. Finally, free valences are denoted by figures at the ends of arrows emanating from the centre of which the free valence is being indicated. (It should be mentioned in passing that the numerical values cited in Fig. 8-1 for aniline are not simple Hückel ones but are obtained from iterative refinements similar to those described in Chapter Seven but taking count, in addition, of the fact that a hetero-conjugated system is being dealt with.) A molecular diagram of the type illustrated in Fig. 8-1 contains a considerable amount of information about the molecule in question. We might, on the basis of such a diagram, therefore ask what sorts of conclusions it would be legitimate to draw, concerning the molecule.

1) Firstly, we note that charge leaves the nitrogen atom—which is, therefore, somewhat more-positive than in neutral nitrogen—and migrates, preferentially, onto the ring carbon-atoms which are *ortho* and *para* with respect to the $—NH_2$ group. There is also a certain amount of negative charge on the carbon atom to which the lone $—NH_2$ group is attached, but charge does *not* go onto the carbon atom which is *meta* to the $—NH_2$ group. This is all as we would expect from the traditional resonance-diagrams. In these terms, there is a certain possibility of a structure in which the nitrogen atom is completely positively charged so that there is a formal double-bond from the $—NH_2$ moiety to the ring. In this circumstance, there are two ways in which the resulting negative-charge can be accommodated in the ring with conventional valence-structures—in the *para* position (Fig. 8-2a) or in the *ortho* position (Fig. 8-2b). What the calculations displayed in the molecular diagram

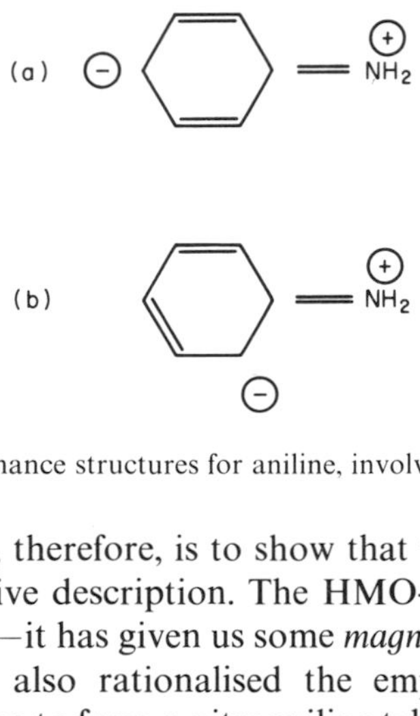

FIG. 8-2 Some resonance structures for aniline, involving charge separation.

(Fig. 8-1) have done, therefore, is to show that this is indeed an entirely appropriate qualitative description. The HMO-treatment has, however, done more than this—it has given us some *magnitudes*, some quantitative information. It has also rationalised the empirical observation that attack by an NO_2^+-ion to form a nitroaniline takes place preferentially in the *ortho*- and *para*-positions—in other words, that the $—NH_2$ group is *ortho*/*para*-directing in an electrophilic reaction. (In addition to these electronic influences, there will of course also be some steric effects in the *ortho*-position but we are not considering these for the present; they may be brought into the discussion as a separate consideration later on (§8.2)).

2) The second point to observe from the molecular diagram of aniline shown in Fig. 8-1 is that the bond orders (and hence, by implication, the bond lengths) are very nearly unaltered from the benzene value. If one were to ask the question to what extent could these bonds really become double bonds (and they are formally, for example, in individual resonance-

structures—Figs. 8-2a and b) the answer would be—very little. The total bond-order of a carbon-carbon bond in benzene is *ca.* 1·67 and the values displayed in aniline are in the range 1·64 – 1·67. Hence, we would not expect any significant changes in bond lengths around the ring.

3) We now consider what information is available from the free valences. The largest free-valence is in the *ortho*-position and hence we would expect this position to be slightly activated in radical reactions, such as attack by a $—CH_3$ radical. The free valence of a carbon atom in benzene is *ca.* 0·40 and the value encountered for the *ortho*-position of aniline is a little larger. Balanced against this, however, in the present example, would be the steric effects from the $—NH_2$ group which an incoming free-radical would encounter on approaching the *ortho*-position.
4) Finally, we note that there is a slight double-bond character about the bond $\rangle C—N \langle$ between the aromatic ring and the $—NH_2$ substituent. A quantitative measure of the C—N bond-order is 1·24—not very large, but enough to show that we are dealing with something more than a pure, single bond.

8.2 Reactivity

We may conveniently divide reactions into those (called *heterolytic*) which primarily involve charges, and those (called *homolytic*) in which charge is not the prime influence.

Calculated π-electron-charges have been very widely used to predict the positions of attack in heterolytic reactions. Let us begin first of all with attack by ions. It was Pauling and Wheland[R28] who originally proposed that a system (represented schematically in Fig. 8-3) containing a positively charged site

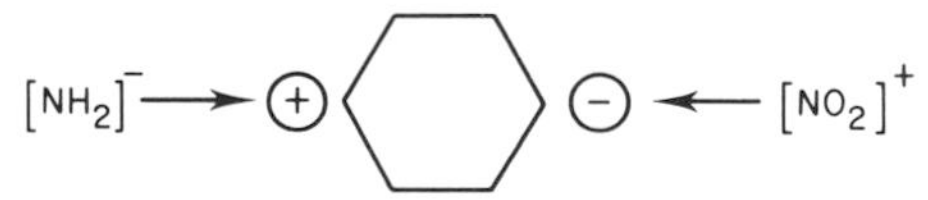

$[NH_2]^- \longrightarrow \oplus$ ⬡ $\ominus \longleftarrow [NO_2]^+$

FIG. 8-3 Schematic representation of heterolytic reactions: (left) nucleophilic and (right) electrophilic.

would undergo preferential attack at that site by a negatively charged group (in, for example, a process of ammination—see the left-hand-side of Fig. 8-3), because of the attraction between unlike charges. Similarly, an ion which is positively charged would preferentially approach a negatively charged centre, as in the process of nitration, schematically represented on the right-hand-side of Fig. 8-3.

If we are interested in the positions in which a particular molecule will react, we should, therefore, first look at the initial charge-distribution; this may givc, at any rate, a clue as to the way in which the reaction in question may proceed. As an example, let us take some rather crudely-calculated π-electron-charges in tropone (Fig. 8-4). This comprises a seven-membered

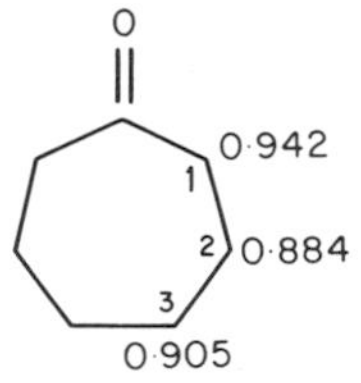

FIG. 8-4 Approximate π-electron-charges in tropone.

ring in which the π-charges on the non-equivalent carbon-atoms are as shown. These are all less than unity for, naturally enough because of its higher electronegativity, the excess π-charges all go onto the oxygen. What we would conclude from an initial inspection of these figures is that a species which was, as-it-were, "looking for" electrons would find that (with the exclusion of the oxygen atom, which is another issue) the place to go would be carbon-atom 1. Even then it is not, on the whole, a good place to seek electron-density for it is less attractive for such an electrophilic attack than, for example, a carbon atom in benzene itself, in which the π-electron charge per carbon atom is unity. On the other hand, a species which was "looking for" (if we may continue to use this highly-anthropomorphic language!) a positive charge would go where there were as few electrons as possible, and that would be the position labelled "2" in Fig. 8-4.

When this sort of analytical process has been carried out, and it has been borne in mind that there may be some extra complications from the carbonyl region, the general shape of the observations initially formulated empirically by Ingold, and Robinson and others[R29], has been theoretically confirmed; but, of course, this kind of discussion has a certain incompleteness about it because it deals only with the situation extant when the reaction is about to begin. An ion approaches the substrate of the molecule and what we are asking is—where will the ion initially be most attracted; where, if one cares to put it this way, will it *start* to move? The answer to these questions may be obtained *via* the sorts of considerations we have just outlined, but the conclusions obtained on that basis do not necessarily mean that the atom will *continue* going towards that particular part of the molecule under attack. For that reason, we require a supplementary discussion which considers, not the beginning of a reaction, but what we might call its transition state. We now therefore discuss what is these days generally called the Wheland intermediate state, or transition state.

Figure 8-5 symbolises some general aromatic system, under attack from

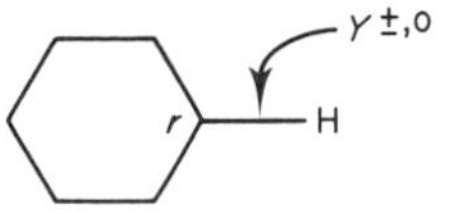

FIG. 8-5 The first stages of an attack by Y^+, Y^- or Y^0 on carbon atom r of a general, aromatic system.

species Y, which could be neutral (Y^0) or it could be Y^+ or Y^-, depending on what kind of reactions are being discussed. Wheland argued that if this atom is imagined to approach the position which is referred to as position r in Fig. 8-5, then there will be a stage in the reaction when $Y^{\pm,0}$ is attached to the rth carbon atom but the proton originally bonded to this carbon atom has not yet left it. This arrangement is illustrated in Fig. 8-6. Naturally,

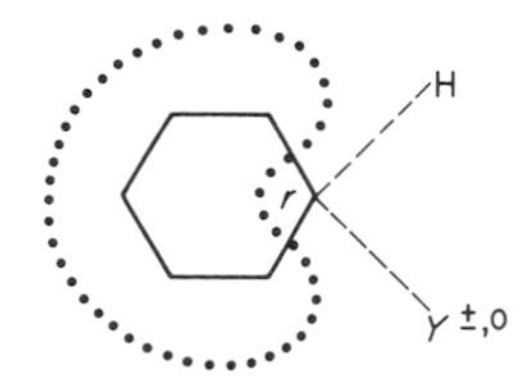

FIG. 8-6 "Residual Molecule" after attack by Y^+,Y^- or Y^0.

for this to happen, the rth carbon-atom must have been converted approximately into a tetrahedral atom—something which would, of course, automatically restrict the range of π-electron mobility. In fact, we could draw a kind of contour which includes within it those carbon-atoms which are still, as-it-were, able to play π-electron rôles. This part of the original molecule is referred to as the *residual molecule*; it is really that part of the molecule which, in this assumed transition-state, is still able to support π-electrons. We have said that we shall suppose that this bond between the rth carbon-atom and Y is properly established, as is that between this carbon atom and H. In order to achieve this arrangement, it is evident that three things involving energy need to have been done.

1) Energy would have been needed to change carbon-atom r from a trigonal state to a tetrahedral one.
2) Energy would have been needed to establish a $C_r - Y$ bond.
3) Some π-electron energy would have been lost as a result of what we might call a limited π-"volume"; by "volume" is meant the extent of the network available to the π-electrons, which is, of course, smaller in the residual molecule than in the original molecule.

Now we should not expect the first two of these to vary very much, either from molecule to molecule or, indeed, from atom to atom within any one, given, molecule. If another carbon-atom (call it s) had been selected instead of the rth one we should find that roughly the same amount of energy was needed to make it tetrahedral instead of trigonal, and the same amount of energy would be needed to establish the $C_s - Y$ bond; this is not *quite* true, but is not too far from the truth. If, therefore, we might assume for the moment that (1) and (2) were constant, then the "activation energy" or "transition energy" should vary directly, and linearly, with (3). We would then fix attention on this third factor, the loss of π-energy consequent upon the more limited π-bonding which may take place in the transition state. We therefore define the localisation energy, E^{π}_{loc}, as

$$-E^{\pi}_{\text{loc}} = E^{\pi}_{\substack{\text{residual}\\ \text{molecule}}} - E^{\pi}_{\substack{\text{original}\\ \text{molecule}}} \tag{8-1}$$

where the quantities on the right-hand-side of equation (8-1) represent the π-electron energies in the residual molecule, and in the original molecule, respectively.

We could at least argue, therefore, that if E^{π}_{loc} were large, so that it was, in a sense, quite "hard work" restricting most of the π-electrons so that they did not get onto C_r, then the transition state would have a high energy. The most-favourable position of attack would be the one in which the localisation energy were the lowest.

Of course, there are *three* sorts of localisation energy, depending upon whether Y, when it comes up, is neutral, or positively or negatively charged; and this will show itself in the number of π-electrons in the residual molecule. If, for example, there are $2n$ electrons in the original molecule, there will be $2n$ electrons if Y is an anion, Y^-, because in this case it comes up with all the electrons needed for the bond C_r—Y and consequently does not require any additional π-electrons in order to form this bond; all the original $2n$ π-electrons are, therefore, "pushed" into the residual molecule. It is evident that there will be $(2n - 1)$ electrons in the residual molecule if Y is Y^0; similarly, there will be $(2n - 2)$ electrons in the residual molecule if Y is Y^+, for the consequence of Y^+ in this model is that both the electrons which are needed for the formation of the C_r—Y bond must be supplied by the π-system. Ultimately, the difference between these three types of reactions could be indicated by the differences in localisation energies arising when the residual molecule has to maintain approximately $2n$, $(2n - 1)$ and $(2n - 2)$ electrons. The first of these is the nucleophilic-, the second the radical-, and the third the electrophilic type of reaction.

Let us consider nitrobenzene as an example. Table 8-1 gives the localisation energies, in units of $|\beta|$, for the three types of attack, nucleophilic, radical and electrophilic, in each of the *ortho*-, *meta*- and *para*-positions.

TABLE 8-1

Position of Attack	$E_{loc}^{\pi\ Electrophilic}$	$E_{loc}^{\pi\ Radical}$	$E_{loc}^{\pi\ Nucleophilic}$
		($\lvert\beta\rvert$ units)	
ortho-	1·886	1·783	1·834
meta-	1·852	1·852	1·852
para-	1·861	1·757	1·809

According to the present arguments, an electrophilic reaction should bring about the lowest localisation-energy when it takes place in the *meta*-position; nucleophilic- and radical attacks are expected to occur preferentially at the *para*-position. If, for example, further nitration were required, it would occur in the *meta*-position—and it would, in fact, take place with only slightly less ease than in unsubstituted benzene itself, for which the corresponding localisation-energy (in the same units) is 1·849. However, it should be said that the difference between 1·849 and 1·852 is so small that it is possibly insignificant. Nucleophilic- and radical attacks, on the other hand, should take place at the *para*-position, and should occur more easily than the corresponding reactions involving benzene.

Now this sort of discussion, which we are here illustrating with just one example, does fit a great many experimental facts; in particular, it enables molecules to be placed in an order of reactivity. Furthermore, in addition to saying whether a reaction is going to occur in the *ortho-*, *meta-* or *para*-positions, one can sometimes actually say whether the reaction is going to proceed more quickly with one molecule than with another.

There is, however, at least one point to be made; we may have overdone the simplification in the above discussion. Let us imagine an NO_2^+-cation coming up to a molecule of the type so far considered (Fig. 8-7). Because

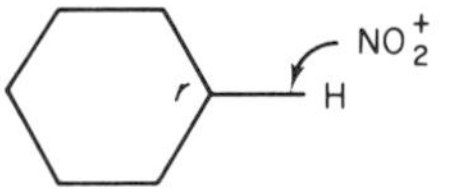

FIG. 8-7 The first stage of an attack by NO_2^+ on the rth carbon-atom of a general aromatic system.

of steric hindrances, it is clearly going to be difficult for the ion to approach in the optimum position for the formation of the transition state since, if the site in question is the carbon-atom, C_r, in the unperturbed molecule, there is a hydrogen atom bonded to it, obstructing the approach of the NO_2^+-ion[N26]; in fact, there are instances when it is almost certainly easier for the NO_2^+ to come down from above, and it is actually known that molecular complexes do exist in precisely that kind of geometrical structure. If, for example, we were to plot a potential-energy surface (or, in this case, a curve) with energy as the ordinate and reaction coordinate as the *abscissa* (Fig. 8-8),

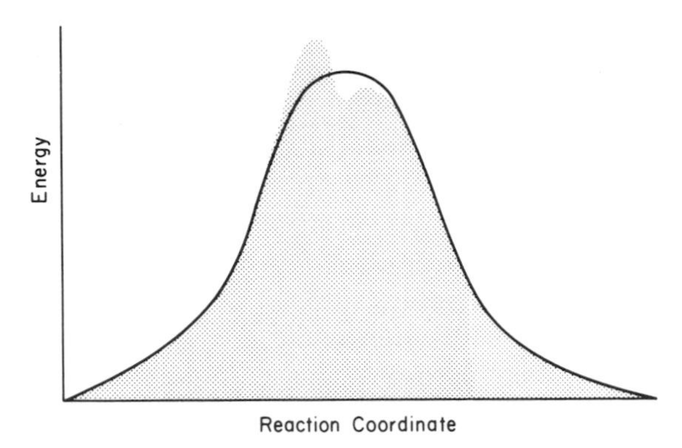

FIG. 8-8 A schematic, potential-energy curve.

and ask whether it is necessarily the case that the peak of this curve is the point at which Wheland's localisation-energies are a minimum, we are forced to conclude that it is not the least bit clear that it is! On the other hand, it might well be, in fact (and the evidence is, in some cases, that it certainly is); the variation of calculated energy with reaction coordinate is then of the form of the dotted line in Fig. 8-8. In this situation, although he may be predicting the correct result, Wheland is not really calculating the activation energy at all. He may be getting something near to it—he may be getting something which varies in the same way; however, the picture that Wheland has provided will have to be regarded as an over-simplification of a given reaction. It is true that this is not always the state of affairs but we have to concede that, until more is known about the geometric description of reactions, it is not really fair to use Wheland's theory too rigorously.

8.3 Differential Relationships Connected with Reactivity

We now examine reactions from a slightly different point of view and introduce certain differential relationships, still considering only the π-electrons. We have already seen (§2.5) that the energy, ε_I, of an electron in an orbital of energy, ε_I, is

$$\varepsilon_I = \sum_u c_{Iu}^2 \alpha_u + 2 \sum_{u<v} c_{Iu} c_{Iv} \beta_{uv} \tag{8-2}$$

The total energy is

$$E^{\pi} = \sum_{I=1}^{n} \nu_I \varepsilon_I \tag{8-3}$$

where ν_I is the number of electrons (2, 1, or 0) in the Ith orbital. Hence

$$\begin{aligned} E^{\pi} &= \sum_{I=1}^{n} \sum_{u=1}^{n} \nu_I c_{Iu}^2 \alpha_u + 2 \sum_{I=1}^{n} \sum_{u<v}^{n} \nu_I c_{Iu} c_{Iv} \beta_{uv} \\ &= \sum_{u} q_u \alpha_u + 2 \sum_{u<v} p_{uv} \beta_{uv} \end{aligned} \tag{8-4}$$

where q_u (equation (4-4)) is the π-electron charge on the uth atom of the conjugated system and p_{uv} (equation (4-6)) is the π-bond order of the bond between atoms u and v. Equation (8-4) is a very interesting and useful relation, for it enables us to obtain some differential properties. We could, for example, imagine changing the Coulomb integral, α_r, on just *one* atom (the rth), and then enquire of the resulting change in E^{π}. Clearly, this change is obtained by differentiating E^{π} partially with respect to α_r, giving

$$\frac{\partial E^{\pi}}{\partial \alpha_r} = q_r \tag{8-5}$$

since the only term in the summation over u in equation (8-4) which survives such a partial differentiation is that in α_r, those in

$$\alpha_u \; (u = 1, 2, 3, \ldots, r-1, r+1, \ldots, n)$$

all being held constant and hence disappearing during the partial-differentiation process. The term involving a summation over u and v in (8-4) is constant for *all* u and v, during a partial differentiation of E^{π} with respect to α_r only, and hence also disappears completely. By an exactly similar argument, for some particular values, r and s, of the running-indices u and v, respectively

$$\frac{\partial E^{\pi}}{\partial \beta_{rs}} = 2p_{rs} \tag{8-6}$$

We see from (8-5) and (8-6) that the charges and bond orders that we introduced in §§4.2 and 4.3 have a kind of fundamental status. We are now able, for example, to alter just one atom in a molecule and perhaps imagine that the effect of this change is to modify only the Coulomb term, possibly by acknowledging that the given atom is now a little more electronegative or a little less electronegative than the atom which it is replacing. This could be achieved by the introduction of some substituent which either pushes electrons towards it or pulls electrons away from it, so that α_r, the Coulomb integral of the atom at which the change in question has been effected, is going to depend upon the charge to be found on the rth atom (see §§7.2 and 7.3). Likewise, if the resonance integral of a given bond (say the r-s bond) is altered—perhaps by altering the length of this bond by some means—equation (8-6) gives information about the consequential changes of energy in the system.

Both the relationships (8-5) and (8-6) have found considerable use. Because we are going to talk more extensively about the first one, (equation (8-5)), we begin with a brief discussion of the second (equation (8-6)). Let us imagine a molecule vibrating, and let us take a very simple case just for the purposes of discussion—benzene, the "breathing frequency" of which is *ca.* 991 cm^{-1}. Now, in the course of this vibration, the bond length is changing—they are all changing equally, that is why we have chosen this example. The bond length is thus fluctuating, and so, therefore, is β_{rs} for this depends upon the bond length (see §7.4); consequently, from equation (8-4), the total π-electron-energy, E^{π}, is also fluctuating. (This, of course, is to be expected; E^{π} ought to get bigger and smaller otherwise we would not have a vibration!) An analysis of this kind is then the basis for calculating the change of energy with bond length, R, (Fig. 8-9) and, consequently, (with a little more care)

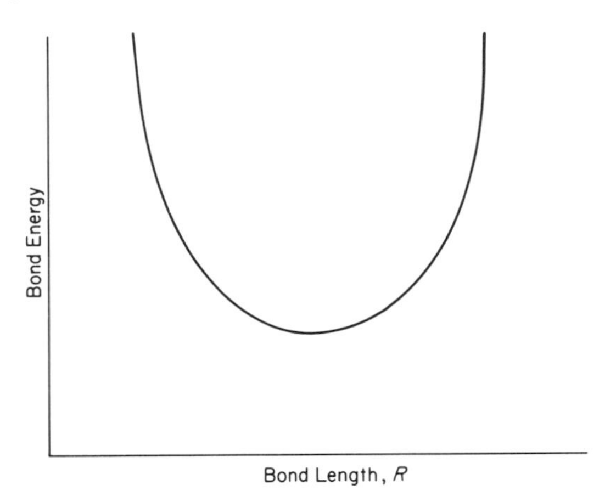

FIG. 8-9 Schematic representation of variation of bond energy with bond length.

of obtaining information about force constants. In benzene there is, of course, very little to do, for this is a very simple example; indeed, benzene is used to calibrate calculations on other molecules—for example, naphthalene.

We now turn to the first of these differential relationships (equation (8-5)), relating the charge of an atom to the rate at which the total π-electron-energy changes on alteration of the Coulomb integral of that atom. This particular relationship enables us to look from a slightly different point of view at some types of chemical reactions.

Let us consider ammination (involving the $—NH_2^-$-ion) or nitration (*via* the NO_2^+-ion) and imagine either one of these ions' approaching one part of a π-electron, conjugated system (Fig. 8-10). We can always assume that the

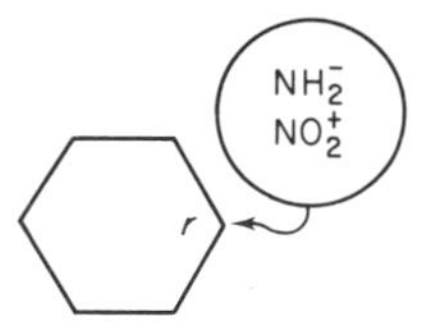

FIG. 8-10 Ammination or nitration at carbon atom r of a general, aromatic system.

π-electron molecule is, say, benzene and we are not going to be too particular, at this stage, about the geometry of the line of approach of these ions to the substrate molecule; we mentioned some of the difficulties about this in §8.2. If the approaching ion were, say, NO_2^+, the electrons in the substrate molecule—and particularly, of course, the π-electrons—will feel a kind of attractive "tug". Furthermore, the "tug" will be greatest towards that carbon atom which is nearest. We could, of course, introduce more sophistication if we wanted to and consider a "tug" or a "pull" on other positions in the substrate molecule. However, the simplest model would be one in which the presence of the positively charged ion in the vicinity of the nearest carbon atom, C_r, is assumed to have the effect of pulling electrons towards C_r—that is to say, of effectively changing the electronegativity of that atom. We might say that the presence of the positively charged ion means that the original Coulomb-integral of the rth carbon-atom, α_r, has to be modified, now becoming $\alpha_r + \delta\alpha_r$. Now whether this $\delta\alpha_r$ is positive or negative is going to depend upon whether it is a positive or negative species which is approaching the substrate molecule.

If we argue in this way, we can, in the early stages of a reaction, forget about the approaching ion and merely discuss the system under attack, *i.e.*, the original molecule, but on the supposition that just one of its Coulomb integrals, α_r, has been modified. Equation (8-4) then describes how the total π-electron-energy changes as a result of this modification, because it will mimic the way in which the π-electron energy changes in the early stages of the reaction. This change, δE^π, will be, to first-order,

$$\delta E^\pi = \frac{\partial E^\pi}{\partial \alpha_r} \delta\alpha_r + O(\delta\alpha_r)^2 \tag{8-7}$$

We can now appeal to our differential relationships (equation (8-5)) in order to convert this into a statement about charges:

$$\delta E^\pi = q_r \delta\alpha_r + O(\delta\alpha_r)^2 \tag{8-8}$$

Now if we may make the reasonable (but-not-exactly-correct) supposition that the σ-bonds behave in much the same way whatever part of the substrate molecule is attacked (they are localised and hence respond to an incoming charged-species in the same sort of way, whatever σ-bond it may be) then the position of the reaction site will be governed entirely by equation (8-8)—*i.e.* will be governed by q_r, since $\delta\alpha_r$ may be thought of as being essentially just the "potential" due to the approaching cation or anion. This suggests that we had a rather less-than-proper argument in §8.2 that, in a heterolytic reaction, the important property determining whether a molecule will be reactive is the charge at a particular centre. Clearly, whether we desire q_r to be large or small for this purpose depends upon whether $\delta\alpha_r$ is positive or negative—in other words, upon whether an electrophilic- or nucleophilic reaction is taking place. We could, if we wished, regard the above argument as a different proof of our earlier claim that it was the charge that mattered.

The reader may, however, object that there are a number of molecules, the alternant hydrocarbons which we discussed in great detail in Chapter Six, in which the charge densities, q_r, at *all* carbon atoms, r, ($r = 1, 2, \ldots, n$) are identically unity, by part 3 of the Coulson–Rushbrooke theorem. Examining the π-electron charge at the various sites in such a molecule is, therefore, no longer a way of distinguishing one position from another. For example, is the α-position in naphthalene more reactive than the β-position, and, if so, why? Such a distinction cannot depend upon the $\{q_r\}$, for, as we have just observed, they are all equal. In that case we shall just have to make appeal to the next-highest-order differential—a procedure which introduces a new set of properties, called polarisabilities, which have proved quite important in the study of this kind of system. The word "polarisability" is rather an unfortunate one, but we shall use it and deal here with so-called "atom-atom polarisabilities".

If we consider expansion of δE^π to the second-order term, we obtain

$$\delta E^\pi = q_r \delta\alpha_r + \frac{1}{2}\frac{\partial^2 E^\pi}{\partial\alpha_r^2}(\delta\alpha_r)^2 + \cdots \tag{8-9}$$

As has been seen, in the case of the alternant hydrocarbons the first term is no longer the discriminating one, for such terms are equal for all r, $r = 1, 2, \ldots, n$; it is, therefore, now the second term that matters.

Let us examine this quantity $\partial^2 E^\pi/\partial\alpha_r^2$ which is the same as $\partial/\partial\alpha_r\,(\partial E^\pi/\partial\alpha_r)$. Now the quantity in brackets, we have already seen, is q_r. Hence:

$$\frac{\partial^2 E^\pi}{\partial\alpha_r^2} = \frac{\partial q_r}{\partial\alpha_r} \tag{8-10}$$

In words, $\partial^2 E^\pi/\partial\alpha_r^2$ measures the rate at which charge is brought onto atom r when the Coulomb term for that atom is altered; this, therefore, is a quantity which is a property of the rth atom of the molecule which gives some information as to the ease, or otherwise, of bringing charge onto that atom; it will be denoted $\pi_{r,r}$.

We can also discuss a quantity, $\pi_{r,s}(r \neq s)$. This is therefore something which concerns *two* atoms and would be a measure of the rate at which the charge on atom r changed when the Coulomb integral, α_s, of atom s was altered—*i.e.*

$$\pi_{r,s} = \frac{\partial q_r}{\partial \alpha_s} \tag{8-11}$$

By use of (8-5) we may write this:

$$\pi_{r,s} = \frac{\partial}{\partial \alpha_s}\left(\frac{\partial E^\pi}{\partial \alpha_r}\right) \tag{8-12}$$

When (8-11) is written this way we can appreciate that the order of the differentiations in (8-12) may legitimately be altered so that we may put

$$\begin{aligned} \pi_{r,s} &= \frac{\partial}{\partial \alpha_r}\left(\frac{\partial E^\pi}{\partial \alpha_s}\right) \\ &= \frac{\partial q_s}{\partial \alpha_r} \end{aligned} \tag{8-13}$$

Hence:

$$\pi_{r,s} = \pi_{s,r} = \frac{\partial q_r}{\partial \alpha_s} = \frac{\partial q_s}{\partial \alpha_r} \tag{8-14}$$

and the relationship is seen to be a *mutual* one. In fact, the quantities, $\pi_{s,r}$, are known as "mutual atom-atom polarisabilities". Although we shall not go into the details of their calculation here, these polarisabilities may be evaluated quite easily, either by use of perturbation theory or, with more sophistication, by integration using simple complex analysis.[R36]

As an example, we refer to naphthalene, mentioned earlier, and enquire whether the α(1)- or the β(2)-position is more reactive in a nitration, ammina-

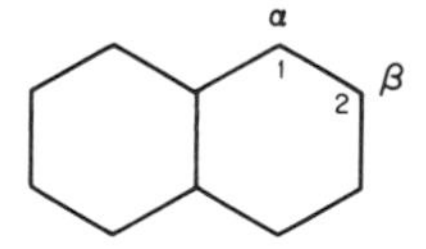

FIG. 8-11 The "α"- and "β"-positions in naphthalene.

tion, etc. (Fig. 8-11). The numerical values of $\pi_{1,1}$ and $\pi_{2,2}$ (which turn out to be in units of $1/\beta$) are

$$\left.\begin{aligned} \pi_{1,1} &= 0{\cdot}443 \\ \pi_{2,2} &= 0{\cdot}405 \end{aligned}\right\} \tag{8-15}$$

These figures lead to the prediction that position 1 is more reactive in electrophilic reactions, as is, in fact, found experimentally.

8.4 The Law of Alternating Polarity

A second illustration which we shall base on naphthalene concerns an empirical law which has become known as the "law of alternating polarity". Let us suppose that we formally convert naphthalene into quinoline. This means that we have modified a particular site in naphthalene because we have replaced a C—H at this particular position by a nitrogen (Fig. 8-12).

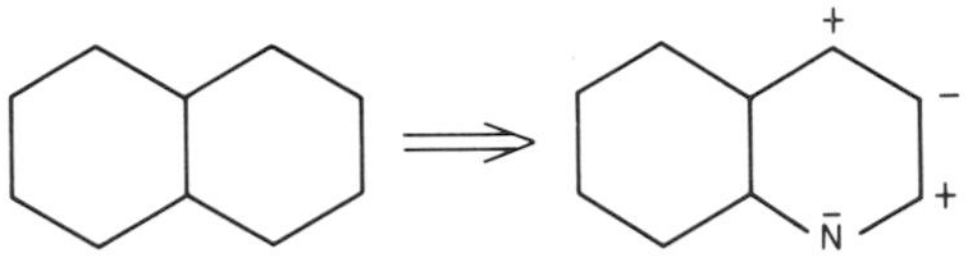

FIG. 8-12 Qualitative consequences of the replacement of a C—H-group in naphthalene by a nitrogen atom.

The law of alternating polarity states that positions increasingly distant from this site of replacement alternately carry greater and less charge than they did in the parent hydrocarbon in which all π-charge-densities were equal. In the present example, the nitrogen atom is more electronegative than the carbons and so draws electronic charge to it. By the law of alternating polarity, the position adjacent to the nitrogen atom in quinoline is therefore (relatively) positive, as indicated in Fig. 8-12, the site next to the positive one, negative, the next one, positive, and so on. The extent of this non-uniformity of charge may diminish and become attenuated with distance from the origin of the perturbation, but its influence on reactivity is immediate, since it implies that certain positions are activated for particular kinds of reactions, while others are predisposed to different types of reactions.

This "law of alternating polarity", which was initially formulated on a purely empirical basis by experimental observation, may be justified by use of some of the coefficients which we have been discussing in this subsection, in particular by $\pi_{r,s}$. We begin by considering a long polyene-chain, schematically represented in Fig. 8-13, and focus attention on two

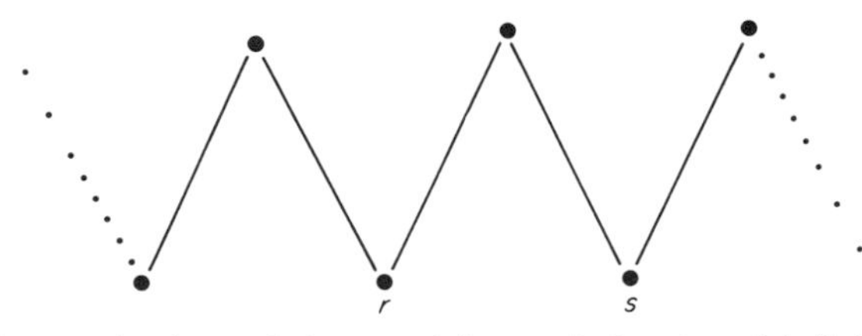

FIG. 8-13 A general polyene-chain, containing particular atoms labelled "r" and "s".

atoms, r and s, in it. Let us imagine that atom r has been altered, its Coulomb integral thus changing from α_r to $\alpha_r + \delta\alpha_r$. Then, as we have seen, the change, δq_s, in the charge on any other atom, say atom s, is simply related to $\delta\alpha_r$ by

$$\delta q_s = \frac{\partial q_s}{\partial \alpha_r}\,\delta\alpha_r + \cdots \tag{8-16}$$

to first order; (higher-order terms may be added if needed). Now $\partial q_s/\partial\alpha_r$ is just one of the mutual atom-atom polarisabilities; therefore, from equation (8-14), we may write:

$$\delta q_s = \pi_{s,r}\,\delta\alpha_r + \cdots \tag{8-17}$$

so that the changes in the π-electron charges on the various atoms along the chain are determined by these mutual atom-atom polarisabilities. In the case of a straight chain it is not difficult to calculate these, and they are, in fact, found to alternate in sign, along the chain. For a straight chain, therefore, (and also for a ring system, providing one does not press too far difficulties concerning the two alternative ways of going from one position in the ring to another) the law of alternating polarity is confirmed by this kind of analysis. Of course, as we have noted, the effects are damped and they die away with increasing distance from the perturbation site—changing sign all the time, however, in accord with the law of alternating polarity.

As a final illustration of these effects we consider some calculations on aminostilbene (Fig. 8-14). We may regard the amino group as providing a perturbation on the end carbon-atom, from the parent, alternant, conjugated hydrocarbon. In Fig. 8-14, beneath the diagram of the molecule, and matching exactly the positions of the various carbon-atoms in it, are plotted the deviations (from the unit electronic-charge applicable to the case of the parent alternant-hydrocarbon) of the charges on these carbon-atoms. Again, this example is seen to illustrate the alternating character of the charge distribution which we have been discussing.

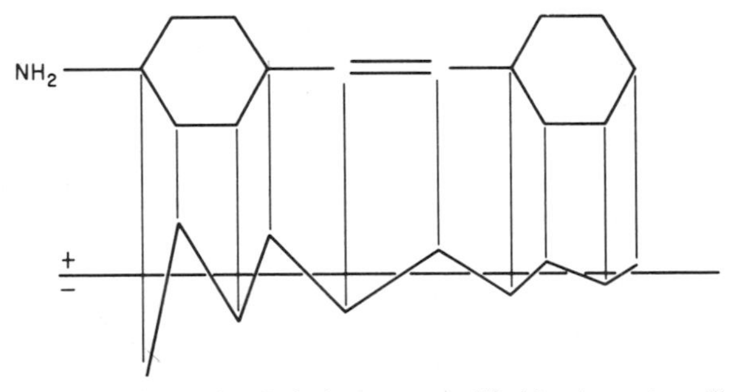

FIG. 8-14 The Law of Alternating Polarity, as exemplified by the aminostilbene molecule.

Nine

Failures of the Hückel Method: An Outline of More Sophisticated Theories

9.1 Introduction

We devote the concluding chapter of this book to a very brief account of the failures of the Hückel method, and of the kinds of ways in which one may seek to remedy them. This really ought to be the opening chapter of a new book but it has, perforce, to be the closing one of the present story. We have already encountered and mentioned various of the deficiencies of the Hückel method as we have progressed through the preceding eight chapters, and these will not be reiterated here. We begin, however, with a topic which we *have* touched on previously—ultra-violet spectra.

9.2 Ultra-Violet Spectra: Lack of Distinction Between Singlet and Triplet States

This is an important phenomenon for any theory to account for, because ultra-violet absorptions of a given molecule are one of the criteria by which molecular structure is routinely tested. Because, in this application, we need to classify a particular absorption, it is important that we should be able to say something about excited states, rather than just the ground states, of molecules. The immediate difficulty which is encountered with the Hückel method is that it does not distinguish states of different multiplicity; (in this instance, the most important examples would be singlets and triplets). Let us suppose, purely for convenience, that we are dealing with a two-electron system (other electrons could also be considered if we wished). In this two-electron system we suppose that one electron is in an orbital, u, and the other is in a different orbital, v—*i.e.*

$$u(1)v(2) \tag{9-1}$$

This may, for example, be regarded as an excited state, in which one electron has been excited from what was initially a doubly-occupied orbital, u, to an initially-vacant orbital, v. Now we know, of course, that this is not an acceptable wave-function, for, as written, it is not anti-symmetrical with respect to the electrons; we must allow for the possibility that electron 2 is in orbital u, and electron 1 is in orbital v. *i.e.*

$$v(1)u(2) \tag{9-2}$$

This means taking both these possibilities into account, either symmetrically:

$$u(1)v(2) + v(1)u(2), \tag{9-3}$$

or antisymmetrically:

$$u(1)v(2) - v(1)v(2) \tag{9-4}$$

In (9-1)–(9-4) $u(i)$, $v(i)$ represent *spacial* orbitals only; the anti-symmetry property of the entire wave-function is then ensured by the product of these space orbitals with parts which depend, in an appropriate manner, on the *spin* ($\alpha(i)$ or $\beta(i)$) of the two electrons. In the case of the symmetrical space-combination ((9-3)), for the whole wave-function to be anti-symmetrical the spin part must itself be anti-symmetrical—*i.e.*

$$\alpha(1)\beta(2) - \beta(1)\alpha(2) \tag{9-5}$$

or, in normalised form,

$$\frac{1}{\sqrt{2}}(\alpha(1)\beta(2) - \beta(1)\alpha(2)) \tag{9-6}$$

In the case of the anti-symmetrical space-combination ((9-4)), there are three possibilities for the required symmetrical spin part—both electrons with α-spin, both with β-spin, or one of each (the latter in a symmetrical, normalised combination)—*i.e.*

$$\alpha(1)\alpha(2)$$

or

$$\beta(1)\beta(2)$$

or

$$\frac{1}{\sqrt{2}}(\alpha(1)\beta(2) + \beta(1)\alpha(2)) \tag{9-7}$$

This is precisely the familiar situation which is encountered in discussions of the helium atom and excited states of the hydrogen molecule in which singlets and triplets differ because the wave functions are not the same, even if the orbitals (*i.e.* the space parts) are.

Now, in the Hückel method, this kind of distinction is not made; we merely assume that we have one electron in orbital u, and one in orbital v; in view of this we would not pose the supplementary question "Do these electrons have parallel or anti-parallel spins?" Consequently, if there is a difference in energy between these two wave-functions (singlet and triplet), that difference in energy must go undetected in Hückel theory. There *is*, in fact, a difference in energy; the expression for it can be written down very easily. Let us suppose that the space orbitals, u and v, are mutually orthogonal; then the difference in energy between the singlet, (^1E), and the triplet, (^3E), (where E refers to the whole molecular-energy) is given by

$$^1E - {}^3E = 2 \int \frac{u(1)v(2)}{r_{12}} v(1)u(2)\mathrm{d}\tau_1\,\mathrm{d}\tau_2 \qquad (9\text{-}8)$$

That is a quantity, by no means negligible, which Hückel theory neglects. In order to give an order-of-magnitude to this neglect, let us consider a simple example, that of ethylene, in which there is a transition which transforms the normal ground-state of the molecule (N) to the triplet state (T):

$$N \Rightarrow T; \qquad {}^1A_{1g} \Rightarrow {}^3B_{1u} \qquad (9\text{-}9)$$
$$\Delta E = 4{\cdot}6 \text{ eV}$$

(In (9-9) we have used group-theoretical terminology).[R31] For this transition, the energy of excitation is 4.6 eV. On the other hand, the singlet transition (generally referred to as "N to V") which starts in a state of $^1A_{1g}$ symmetry and ends in state $^1B_{1u}$

$$N \Rightarrow V; \qquad {}^1A_{1g} \Rightarrow {}^1B_{1u} \qquad (9\text{-}10)$$
$$\Delta E = 7{\cdot}7 \text{ eV}.$$

has an excitation energy of 7·7 eV. Thus

$$^1E - {}^3E = (7{\cdot}7 - 4{\cdot}6)\text{eV} = 3{\cdot}1 \text{ eV} \qquad (9\text{-}11)$$

This is to say that the difference between the singlet and the triplet—which involves, in both cases, precisely the same *space orbitals*, u and v—is 3·1 eV. Now 3·1 eV, when we are only dealing with about twice this amount of energy in all, is a rather sizeable proportion of the whole! Very little sense, therefore, will be made of the interpretation of ultra-violet spectra if Hückel theory is used in situations where singlet- and triplet states need to be distinguished.

The above is, perhaps, an extreme example, but it does indicate the caution which must be exercised. In order to obviate difficulties of this kind, people have sometimes rather shamelessly chosen one value of β when dealing with singlet transitions, and another value of β when dealing with triplet transitions! This is, perhaps, permissible in certain cases, but, in adopting this policy, one should not disguise from oneself the fact that justice has not been done to the difference in multiplicity between the singlet and triplet forms.

These considerations are, of course, really bound up with the Hückel theory's failure to deal with the electron-interaction terms which arise— $1/r_{ij}$ for each pair of electrons, i and j. In the Hückel scheme, an attempt is made to average such terms in order to obtain the effective, one-electron Hamiltonian, $\mathscr{H}_{\text{effective}}$ (§2.2). It is during this averaging process that the singlet-triplet distinctions which we have been describing in this subsection are lost.

9.3 The Extended-Hückel Method

In the Hückel theory as we have been using it we have, effectively, taken no account of the presence of the σ-electrons. In the early days, people would have said, without any hesitation, that the σ-electrons must be treated in the same way as the π-electrons, with, perhaps, some interaction between the two. That dictum became completely forgotten until it was revived relatively recently and given the name "Extended Hückel".[R32]

The philosophy of the Extended-Hückel method is that one must use an HMO-theory for the σ-electrons as well as the π. Incidentally, at the same time, we may add some overlap-integrals if we think it is the appropriate time to do so. In the case of σ-electrons, this may be more important than it is for the π because σ-orbitals tend to overlap rather more effectively than the π-orbitals; that is why, on the whole, they form stronger bonds. The σ- and π-electrons will exert some influence on each other, as the following simple analysis illustrates. Consider two atoms, A and B, in a molecule (Fig. 9-1);

FIG. 9-1 The two atoms of a diatomic molecule, A—B.

let us assume that there is a concentration of π-electrons at A, so that the charge, q_A^π, is large. The fact that there is an excess of π-electrons on A will cause σ-electrons, on sheer electrostatic grounds, to be pushed away from A. This will therefore influence the σ-electron Coulomb-term for atom A (which we shall call $\alpha_A^{(\sigma)}$, to distinguish it from the familiar π-electron Coulomb integral, α_A, here temporarily called $\alpha_A^{(\pi)}$). Immediately, therefore, we begin to see how the σ- and π-electrons will influence one another; this has important consequences for a number of recently measurable phenomena, particularly those involving excited states. Suppose, for example, that we have been able to calculate the σ-electron and π-electron dipole-moment of a given molecule in its ground state. If we then consider a π-electron excitation from an orbital, π, to one which we shall label π^*, *i.e.*

$$\pi \to \pi^* \tag{9-12}$$

it is rather unlikely that the charge distribution in the π^*-orbital will be the same as that in the π-orbital from which the transition started. It could happen that it was and, as we have seen (§6.5), the long-wave-length absorptions which involve transitions from the highest-occupied (HOMO) to the lowest-unoccupied (LUMO) orbitals will frequently be of this sort. But, in the case of other transitions, this will not be so; hence, in general, the π-electron charge-distribution, $\{q_r^\pi\}$, $r = 1, 2, \ldots, n$, will change, upon such an excitation. There will be a consequential throw-back onto the σ-electrons, as our example involving Fig. 9-1 illustrated. This means that $\{\alpha_r^{(\sigma)}\}$, $r = 1, 2, \ldots, n$, the Coulomb-integral set for the σ-electrons, will change; so also will the charge distribution, $\{q_r^\sigma\}$, $r = 1, 2, \ldots, n$. Immediately, therefore, we have, in principle, a way of dealing with this mutual interaction between the electrons in these two sets of symmetrically distinct types of orbital. This is an important step forward for there are now ways in which dipole moments in excited states may be measured; furthermore, it is found that if calculations of molecular dipole-moments are carried out on the supposition that the σ-framework of the molecule in question remains inviolable, the predicted dipole-moments are far, far too big. Thus, for example, a member of Coulson's group, R. F. Weaver[R33], calculated the π-electron dipole-moment of a certain excited-state species and found it to be 15 Debye-units in one particular direction; when σ-electron interactions were taken into account, however, the σ-electron contribution to the dipole moment was found to be -14 Debyes, giving, as a resultant, quite a modest dipole-moment, typical of that experimentally encountered in excited species of the sort considered.

This method is very much in its development stage at present, but if one wished to consider both σ- and π-electrons in a calculation of the type we have been discussing in this book, the Extended-Hückel approach represents one way in which this could be done.

9.4 The Pariser–Parr–Pople Method and Roothaan's Equations

The Extended-Hückel method is still based on the old idea of an effective Hamiltonian (§2.2) in which the $1/r_{ij}$-terms have been averaged and it will, therefore, still have the limitations which we have already considered (in

§9.2) in the context of the simple HMO-treatments of the π-electrons alone. Since the discussion of §9.2 has shown that we really need to include these $1/r_{ij}$-terms in some way, we next turn our attention to a method, simultaneously developed by Pariser and Parr[R35], and by Pople[R35], which accomplishes this, The PPP-method (as it is almost universally acronymed) still deals with π-electrons only, but it goes beyond the Hückel theory in that it uses a proper π-electron-Hamiltonian which explicitly includes the $1/r_{ij}$-terms, involving π-electrons i and j; however, since the PPP-method considers only a π-electron Hamiltonian, $1/r_{ij}$-terms which describe interactions between a σ-electron and a π-electron do *not* appear explicitly and still, therefore, have to be taken into account in some sort of average way. In fact, the σ-electrons and the nuclei of the molecule are considered to provide a field (or "core") in which the behaviour of the π-electrons may be studied.

As we have just seen, the PPP-method does properly take care of $1/r_{ij}$-interactions between pairs of π-electrons. This philosophy had in fact also been approached by Roothaan[R36]; (the publication sequence is a little confusing because Roothaan's work was done about ten years before he published it and, in the meantime, various other people had also made attempts at the problem). Roothaan formulated a set of equations which have become known as "Roothaan's Equations". These are a generalisation to this full π-electron-Hamiltonian of the secular equations which we encountered (during Chapter Two) in Hückel theory. In §2.3, we wrote the secular determinant arising in Hückel theory in the form:

$$|H_{rs} - \varepsilon S_{rs}| = 0 \tag{9-13}$$

H_{rs} was the matrix-component of the Hückel effective-Hamiltonian operator, $\mathscr{H}_{\text{effective}}$, between two basis atomic-orbitals, ϕ_r and ϕ_s, S_{rs} was the overlap integral between ϕ_r and ϕ_s, and H_{rr} was set equal to α, H_{rs} to β. This is how we developed the simple HMO-approach in Chapter Two. What Roothaan did was to show that a formally similar determinant is obtained in a full treatment of the π-electrons, but that it involves a somewhat more complicated expression for the matrix-elements, H_{rs}. Furthermore, he showed that this more-complicated expression somehow had to take into account interactions between any one π-electron and all the other π-electrons. We do not go into the details of this here, except to say that, in order to find the LCAO-MO coefficients for *one* molecular orbital, it is necessary to know all the others, because all the others appear in the expressions for the equivalent terms, H_{rs}. This is a very familiar situation which mathematicians have long known how to deal with and which we encountered during our discussion of the "self-consistent" Hückel-methods in §§7.2–7.5; it is necessary to use an *iterative* scheme. An initial guess is made of all the orbitals except one and these are used to calculate the H_{rs}-terms for the one orbital which has not yet been dealt with; then one of the others is missed out and all the other "guessed" orbitals, plus the one for which the H_{rs}-terms have just been estimated, are used to improve this second orbital. So the refinement continues until further iteration effectively causes no change in the calculated orbitals, between one iteration cycle and other; the iteration has then converged. This was the way in which Hartree performed his self-consistent-field calculations for atoms, a parallel usage.[R37, R38]

Such an iterative procedure does, of course, involve considerably more computational effort. These days this can be conveniently and routinely handled by computers and the extra labour involved is certainly rewarded by the acquisition of better results. Furthermore, if one is prepared to make some rather drastic approximations, a great deal of simplification can be achieved. Pariser, Parr and Pople[R34, R35] decided to neglect almost all integrals involving two distinct atomic-orbitals. They did not abolish them all, but left in ones such as

$$\int \phi_r^2(1) \frac{1}{r_{12}} \phi_s^2(2) \mathrm{d}\tau_1 \, \mathrm{d}\tau_2 \tag{9-14}$$

The reader may feel that this was a rather wicked thing to do! However, it can be shown that, in many cases, this policy gives quite good results; if it is combined with a wise choice of the various parameters which arise in the calculation (a process which is now known as "parametrisation"), it is possible to obtain much information about excited states, even though the procedure has involved neglecting a large number (in fact, by far the greatest number) of the integrals that occur.

9.5 Methods Involving Neglect-of-Differential-Overlap

Having shown that the above procedures work reasonably for the π-electrons, it is natural to go back to the form of equation (9-13) and enquire whether we cannot cease restricting ourselves to π-electrons and start taking the σ-electrons into account in the same way. We cannot, however, discard all the inconvenient integrals in a haphazard and casual manner; some definite set of rules must be adopted to form a consistent policy for deciding which integrals are to be retained and evaluated, and which are to be neglected, in any given calculation. This has resulted in several models, the main one being the CNDO (Complete-Neglect-of-Differential-Overlap) method. There are even several different versions of this, referred to as CNDO/1, CNDO/2, CNDO/BW, etc. Then there is something less severe—INDO (Intermediate-Neglect-of-Differential-Overlap). There are also others, for example, MINDO (Modified-Intermediate-Neglect-of-Differential-Overlap).[R39, N27]

9.6 Calculations based on Gaussian Atomic-Orbitals

The main difficulty in all these methods is the evaluation of so-called "four-centre" integrals, *e.g.*

$$\int \phi_A(1)\phi_B(2)\frac{1}{r_{12}}\phi_C(1)\phi_D(2)\mathrm{d}\tau_1\,\mathrm{d}\tau_2 \tag{9-15}$$

which involve distances measured from four different nuclei; there are very few ways of coping with an integral of this sort—they involve thoroughly difficult mathematical techniques. People have, therefore, looked for some quite different way with which, perhaps, to use rather different functions. The reader will recall that, so far, we have talked about the atomic orbitals as if they were precisely the atomic orbitals to be found on an isolated atom; they might be modified a little by changes of screening constant[R40] but, basically, they were the same. A 1*s*-orbital was a 1*s*-orbital whether it was in an atom or a molecule, and 2*s*- and 2*p*-orbitals were similarly the same in the molecule, as in the atom. It was, however, noted by the late S. F. Boys in Cambridge[R41] that, if we were prepared to do so, we might choose a totally different sort of function, from the Slater functions[R40] (or atomic orbitals) that we have previously used. These are the so-called "Gaussian" functions.

Suppose, for example, that we wish to describe the wave function of an electron at point, *P*, relative to a nucleus at point *A* (Fig. 9-2) (*P* is to be

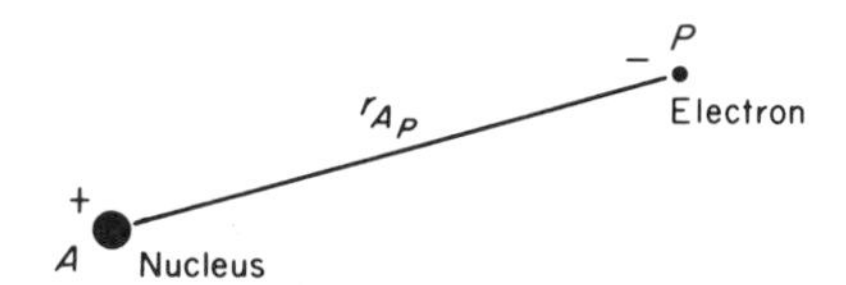

FIG. 9-2 An electron at point *P*, related to a nucleus at *A* by the vector $\mathbf{r}_{AP}$.

thought of as associated with this particular nucleus—it may be, for example; in an *s*- or a *p*-orbital). Then, Boys observed[R41] that if, instead of using the conventional Slater-orbital form[R40] such as

$$\mathrm{e}^{-\zeta r_{AP}}, \tag{9-16}$$

a function of the type

$$\mathrm{e}^{-kr_{AP}^2} \tag{9-17}$$

were employed, every integral which occurs may be evaluated in closed form. In practice, we use not just one of these "Gaussian" functions, but several of them. We might, for example, include a term such as

$$r_{AP}^n\mathrm{e}^{-kr_{AP}^2} \tag{9-18}$$

and, instead of a single term like (9-18), take a sum of several of them. Individually, terms such as (9-17) and (9-18) are inadequate since they do not correctly simulate the behaviour of an orbital at large distances from the nucleus on which it is centred. If, however, we took a linear combination[R42] of, say, three, four, or five of them, we might, by adding them together in this way, obtain something which authentically mimicked the conventional, exponential-type of Slater orbital ((9-16)).

This is the greatest development of the last ten years; people have used combinations of Gaussian functions such as (9-17) and (9-18) instead of the traditional exponential-types ((9-16)). The revolutionary consequence of this is that it is no longer necessary to make the approximation of neglecting all the integrals which previously were difficult or impossible to evaluate. There is, however, one snag. As we have observed, in order to represent a function such as (9-16) by a combination of functions like (9-17) and (9-18), it is found, from experience, that about five such Gaussian functions are required. This means that wherever we had one orbital before we have a combination of five now. Since many of the integrals involved are four-centre ones of the type shown in (9-15), this is quite serious for if, for example, each $\phi_x(i)$ in (9-15) were simulated by a linear combination of five Gaussian functions, there would actually be 625 individual contributions to an integral of that sort! It is not surprising, therefore, that a large molecule like pyridine, when treated in this way, requires the evaluation of 3×10^9 integrals[N28]. Even a calculation (involving, altogether, 90 Gaussians) on something as small as ethane required seven million integrals!

9.7 Note on the Abandonment of σ-π Separability in Non-Empirical Calculations: Configuration Interaction

All the discussion of the previous eight chapters of this book has been based on the model of σ-π separability (§1.2). The grounds for this separability were that π-electron orbitals have a different symmetry with respect to the molecular plane from the σ-electron orbitals (§1.2). The physical picture that emerged was shown to have great power in explaining many characteristic features of π-electron chemistry. It was also exceedingly pictorial, so that we could almost "feel" what was happening in a large, conjugated molecule. But, of course, such a model is not, and never can be, accurate. There are Coulomb- and exchange interactions between σ and π electrons which cannot adequately be described by an equivalent potential; as we have seen (§9.3), displacements of σ-electrons will inevitably lead to consequent displacements (often in the opposite direction) of π-electrons, and *vice-versa*. On several occasions in earlier parts of this chapter we have suggested that a better policy would be to abandon the strict σ-π separability and calculate both σ- and π-MO's simultaneously. However, if we retain one, single, electronic configuration, this may be represented as one single, large, determinantal wave-function comprising both a space-("orbital") and spin part, and each MO will still have either σ or π character[R43]. If, however, we proceed to the natural extension of this and include an interaction between all possible electronic configurations—*i.e.* all possible ways in which available electrons can be placed in the MO's, taking account of their spin—a technique now known as "configuration interaction", all that we ask of the various configurations that are to be mixed together in a linear combination is that each of them shall have the same overall symmetry. They may achieve this even with different numbers of purely π- and σ-MO's. Thus, so far as the total σ-π-symmetry is concerned, σ^2 and π^2 have the same symmetry, as do $\sigma^n\pi^m$ and $\sigma^{n-2}\pi^{m+2}$. When we have reached this stage, the sharp distinction on which the first eight chapters of this book were based disappears. Unfortunately, with this there also disappears the possibility of any simple, pictorial description of most of the characteristic experimental features we have been discussing! Because of the tremendous amount of extra labour involved, however, relatively few calculations have been made in which anything like a full configurational-interaction, including σ- and π-MO's, is incorporated. Most calculations continue to be based on a single-determinantal wave-function, or perhaps a combination of just a few such determinants, in which the distinction between σ- and π-electrons still survives.

9.8 Some Typical Results of Less-Approximate Calculations

(a) Calculations on Pyridine

We conclude by considering one or two results which may be obtained from more-sophisticated calculations of the sort which have been briefly outlined in the last few subsections. We begin by taking pyridine (Fig. 3-3a) as an example. The type of electron which is most easily removed is a π-electron; let us therefore consider ionising a π-electron in the uppermost occupied-orbital. Before the ionisation has been effected, there was, on average, one π-electron *per* nucleus for each of the six nuclei in the conjugated ring. There will, however, be consequential changes which occur as a result of this ionisation process, and it is on these changes that we are going to focus attention. In particular, there will be some relaxation which will cause redistribution of the electrons and, consequently, some shift of charge. It is found[R42], from calculations of the sort we have described, that approximately 0·22 units of π-electronic charge is caused to move from one atom to another, within the pyridine ring, during this ionisation process. Astonishingly, however, an accompanying change of 1·3 units of σ-electron charge takes place at the same time. This seems to imply that taking away a π-electron affects the σ-electrons much more than the π! Well, that is a vivid reminder of how important it is to take account of the σ-electrons, as well as the π.

(b) Calculations on Pyrrole

The second example we consider is pyrrole (Fig. 3-3b), a molecule which, in group-theoretical terminology, has C_{2v} symmetry. Of the fifteen σ-orbitals, nine are of type a_1 and six are of type b_2. Of the π-orbitals, there are two of type b_1, and there is one of type a_2, giving a total of 3 for the occupied π-orbitals. The sum-total is therefore 18. It is thus quite evident that by going from a consideration of π-electrons only (as *per* Hückel) to this more-complete treatment, we have increased the amount of work involved enormously, for the number of molecular orbitals which have to be handled has increased from three to 18. Let us specialise to consider the nitrogen atom in pyrrole and work out how many *s*- and *p*-electrons are associated with it. This turns out to be[R42, R43]

$$1s^2\ 2s^{1\cdot37}\ 2p\sigma^{2\cdot38}\ 2p\pi^{1\cdot66}$$

(The two electrons in the 1*s*-orbital may, of course, be regarded as being in an "inner-shell" and are, as such, outside our present considerations). Here (as in §1.2), $2p\sigma$ denotes a *p*-orbital which is symmetrical with respect to reflection in the plane of the molecule. In the isolated nitrogen atom, the configuration is

$$1s^2\ 2s^2\ 2p\sigma^1\ 2p\pi^2$$

We see, therefore, that when the nitrogen atom has become a part of the pyrrole molecule it has lost π-electrons to the extent of 0·34. It has, on the other hand, gained σ-electrons—to the extent of 0·75 overall. The nitrogen atom has thus made a net gain of 0·41 electrons. By these kinds of calculations, we can therefore begin to talk in considerably more detail about the behaviour of an atom in a molecule, and the rôle which it plays in any particular molecule.

(c) Hyperconjugation

The third example concerns hyperconjugation[N29]. We therefore consider a methyl group attached to some conjugated system—let us consider methyl-

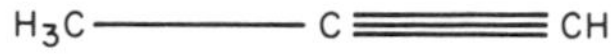

$H_3C—C≡CH$

FIG. 9-3 Structural formula of methyl-acetylene.

acetylene (Fig. 9-3). Mulliken then suggests treating the three carbon-bonds to the hydrogens of the methyl-group as a pseudo triple-bond; we really then effectively have a single carbon-carbon bond lying between two multiple bonds, and this is the arrangement which is called hyperconjugation. It is not "ordinary" conjugation because the H—C (with two further H atoms attached to the C) is only a *pseudo* triple-bond; hence the name, *hyper*conjugation. We know, of course, from general chemical observations, that the methyl group substituted at various sites has significant effects on molecules, particularly in terms of their reactivity—that, in fact, was how Baker and Nathan first discovered it[R44]. Furthermore, much discussion has taken place over many years as to the relative importance of the two parts of this whole phenomenon; one is the purely conjugative part in which the (pseudo) π-electrons can migrate into and out of the rest of the (genuinely conjugated) molecule, and the other part is an inductive effect, in which the methyl group is supposed to alter the nature of the carbon atom to which it is attached. In the jargon of organic chemistry we might say that an electron-donating substituent pushes electrons onto the carbon atom to which it is attached, thereby changing the electronegativity of that carbon atom.

Now it is only within the last few years that it has been possible really to sort out quantitatively these two aspects—inductive and conjugative—of what, collectively, we would call hyperconjugation due to an alkyl (*e.g.* methyl) substitution. Again, resolution of this problem involves unscrambling the σ- and π-behaviour. The π-behaviour gives rise to what we might call the delocalisation effect, because, as a result of this pseudo-conjugation, the "volume" available to the delocalised π-electron system has been extended. The σ-behaviour, on the other hand, can be thought of as being responsible for the inductive part. Now, for the first time, numerical values are available; in fact, calculation shows[R45] that the methyl group donates 0·055 σ-electrons and 0·028 π-electrons to the triple bond. (In this simple example, of course, the triple bond alone constitutes the entire, conjugated, π-electron system). It seems, therefore, that the σ-donation, which gives rise to the inductive effect, is numerically greater than the π-donation, which is normally implied by the word "hyperconjugation". Closer examination, however, shows that the inductive effect dies away with distance very much more rapidly than does the π-electron delocalisation-effect. This is so since the π-electron effect is concerned with delocalised orbitals which extend over the entire framework of the molecule in question; if, for example, the molecule were toluene, the *ortho*-position, and even the distant *para*-position, would be activated as a result of the methyl π-electron-donation. The greater mobility of the π-electrons thus enables the conjugative effect to be transmitted to more-distant parts of the molecule. The fact that it is what would seem to be the smaller of the two contributions which, at larger distances from the substitution site, is in fact the more important, has given rise to some difficulty, in the minds of many people, in understanding this phenomenon of hyperconjugation. The complete phenomenon is, however, now explicable—but its elucidation requires simultaneous study of both σ- and π-molecular orbitals.

We may say, as a final word of caution, that even the more-sophisticated kinds of calculation discussed in this subsection are not to be treated as Gospel-truth, for the figures cited above are valid only within the approximations relevant to the particular methods used to calculate them. The point we wish to emphasise, however, is that, without these methods, we really would not have any understanding of phenomena such as hyperconjugation. The reader may feel that, in this subsection, we have been alternatively an optimist and a pessimist: this would be fair comment and one can only answer that, in this game, one has to be![N30]

9.9 Final Comments on Hückel Theory

In our detailed, and fairly leisurely, excursion into Hückel theory, we have found that it can often give results, such as charges or bond orders (Chapter Four), in which very few (and sometimes no) unknown parameters are involved. This is also true of the energy-level patterns discussed in Chapters Five and Six. Where parameters *are* needed it is seldom very wise to calculate them *ab initio*; it is much better to choose them so that certain properties for one or two molecules are correctly fitted, and then to carry on with those values for other molecules. In this sense it is very much a semi-empirical theory since it relies on experimental data for the quantification of its parameters. If we are looking for general properties of a system of similar molecules, such as the homologous series, benzene, naphthalene, anthracene, tetracene, . . . , it will frequently give us valuable information about the way in which these properties change along the series. It can indicate, both qualitatively and semi-quantitatively, the changes in a molecule that result from a substitution, and it can help to rationalise a large variety of chemical reactions (Chapter Eight).

As we have seen vividly in the present chapter, however, the technique of Hückel theory does have its limitations. These are particularly evident when dealing with excited states (§9.2), where, although it correctly characterises the symmetry properties, Hückel theory cannot distinguish singlets from triplets arising from wave functions which have the same MO's but different spin-relationships. Moreover, there is no reason for the fundamental parameters', α and β, having the same values for the calculation of different properties. Finally, true self-consistency is difficult to achieve (Chapter Seven) and, indeed, cannot be achieved without some explicit introduction of the σ-core (Chapter Nine). Thus, not only are there Coulomb and exchange interactions between the σ- and π-electrons, but any polarity arising from the π-electron distribution will inevitably influence the σ-electron distribution (§9.7).

We conclude, therefore, that the merits of the Hückel scheme lie in its great simplicity, and its efficacy as a kind of interpolatory procedure. Moreover, the chemical insight that it provides is of immense value, both in itself, and as a guide to the more-complete calculations, some of which we have briefly reviewed in the present chapter. In other words, we must accept that Hückel theory does the right job very well—but that, for accurate, numerical values, something vastly more elaborate is needed.

Appendix A

Some Graph-Theoretical Aspects of Hückel Theory

For the benefit of the more mathematically-inclined reader we can conveniently illustrate here the relevance of abstract *Graph Theory* (a branch of Pure Mathematics) to simple Hückel molecular-orbital theory.

In graph theory a graph is defined as a set of points (called *vertices*) some of which are joined to each other by lines (called *edges*). This is a somewhat intuitive and heuristic definition which could be made more formal but it will suffice here for all we wish is to give the reader the "flavour" of the subject. An example of such a graph is the diagram used schematically to represent the connectivity of the carbon atoms in butadiene in Fig. 2-6c. Here the vertices correspond to carbon atoms whilst the edges can be taken to represent carbon-carbon σ-bonds. If the points (vertices) are numbered (as is the case with the graph of Fig. 2-6c), the resulting graph is called a *labelled graph*. All graphs which represent the σ-bond connectivities of the atoms involved in conjugated systems will be *simple, connected graphs* in which there is never more than one edge joining a given pair of vertices and in which each vertex is joined to at least one other. In this situation, one can define the *vertex adjacency-matrix*, $\mathbb{A}$, of such a graph with n vertices, as the matrix $[a_{ij}]$ which has elements

$$a_{ij}(i \neq j) = \left.\begin{array}{l} 1 \text{ if vertices } i \text{ and } j \text{ are} \\ \quad \text{joined by an edge} \\ = 0, \text{ otherwise} \end{array}\right\} \tag{A1}$$

and in which the diagonal elements, a_{ii}, are also zero. The adjacency matrix, $\mathbb{A}$ (butadiene), of the graph in Fig. 2-6c, for example, as labelled in that figure, is thus

$$\mathbb{A}\text{ (butadiene)} = \begin{pmatrix} 0 & 1 & 0 & 0 \\ 1 & 0 & 1 & 0 \\ 0 & 1 & 0 & 1 \\ 0 & 0 & 1 & 0 \end{pmatrix} \tag{A2}$$

The eigenvalues of $\mathbb{A}$ are, by definition, the roots of

$$|\mathbb{A} - \lambda \mathbb{1}_{n \times n}| = 0 \tag{A3}$$

in which $\mathbb{1}_{n \times n}$ is the unit matrix of dimension n. In the case of the butadiene example, by substituting $\mathbb{A}$ (butadiene) (equation (A2)) for $\mathbb{A}$ in (A3), this latter equation becomes:

$$\begin{vmatrix} -\lambda & 1 & 0 & 0 \\ 1 & -\lambda & 1 & 0 \\ 0 & 1 & -\lambda & 1 \\ 0 & 0 & 1 & -\lambda \end{vmatrix} = 0 \tag{A4}$$

Comparison of this with the form of (2-58) shows that when energies are expressed relative to α, and in units of β, the Hückel Hamiltonian-matrix for a given conjugated-hydrocarbon is isomorphic with (and, in the zero-overlap approximation, identical to) the vertex adjacency-matrix of the correspondingly labelled graph. This correspondingly labelled graph represents the σ-bond connectivity of the atoms comprising the conjugated system in question. Since, when energies are expressed relative to α, and in units of β, $\mathbb{A} = \mathbb{H}$ in the zero overlap case, we have that $\{\lambda_I\}$, $I = 1, 2, \ldots, n$, the eigenvalues of $\mathbb{A}$ (the roots of (A3)) are *identical* to $\{x_I\}$, $I = 1, 2, \ldots, n$, the roots of the appropriate secular-determinant (its form will be similar to the one in equation (2-58)). In addition, as has been emphasised, the MO-energies, $\{\varepsilon_I\}$, $I = 1, 2, \ldots, n$, are simply and directly related to the $\{x_I\}$, *via* the (constant and empirical) parameters, α and β, as in (2-59).

In conclusion, since $\mathbb{H}$ and $\mathbb{A}$ commute (*i.e.*,

$$\mathbb{H}\mathbb{A} - \mathbb{A}\mathbb{H} = \mathbb{O} \tag{A5}$$),

they also have the same *eigenvectors*; but the eigenvector of $\mathbb{H}$, $\{c_{Ir}\}$, $r = 1, 2, \ldots, n$, corresponding to the eigenvalue, x_I, is the set of LCAO combinatorial-coefficients associated with the Ith MO, and hence these also are a function only of $\mathbb{A}$, and thus only of the carbon-atom connectivity of the molecule in question.

The above observations emphasise once more that in the simple HMO-approach for hydrocarbons,

1) the MO energy-levels and,
2) the LCAO combinatorial-coefficients,

are determined *solely and entirely* (to within the parameters α and β) *by the carbon-atom connectivity*—this is often (but unfortunately) called the "topology"—*of the conjugated system under investigation.*

For further details of the graph-theoretical aspects of HMO theory, see I. Gutman and N. Trinajstić, "Graph Theory and Molecular Orbitals", *Fortschritte der Chemischen Forschung* (*Topics in Current Chemistry*), **42**, 49 (1973), and D. H. Rouvray, "The Topological Matrix in Quantum Chemistry", Chapter 7 of *Chemical Applications of Graph Theory* (Editor: A. T. Balaban), Academic Press, London, 1976.

R. B. Mallion and B. O'Leary

Appendix B

Application of Group Theory in Hückel Theory[R46]

B1 Introduction

When the π-electron framework of a molecule consists of n atoms, the secular determinant will be of order $n \times n$. Frequently, the framework will possess certain symmetries, such as reflections or rotations; their presence is disguised in the full secular-determinant. By systematic use of such symmetry-properties as exist, not only may the full secular-determinant be factorised into independent sub-determinants of lower order, but much more insight may be obtained into the nature of the MO's. The symmetry properties are independent of the way in which we approximate the MO's, and so the information provided by group-theoretical techniques has a significance far greater than that of any particular set of approximate MO's.

The symmetry properties with which we are concerned arise from the symmetry of the Hamiltonian; in Hückel theory, this means the symmetry of the effective, one-electron Hamiltonian (§2.2). Clearly, this has the same symmetry as the molecular framework itself.

B2 An Example: Naphthalene

We illustrate these ideas by choosing, as an example, naphthalene. In Fig. B1 the x-direction lies perpendicular to the molecular plane; the ten atomic $2p_\pi$-orbitals are therefore of the p_z-type. The molecular framework, and hence also the one-electron Hamiltonian for this molecule, has the symmetry D_{2h}. The x-, y- and z-axes are all two-fold rotation-axes, and the three coordinate-planes are all reflection-planes of symmetry. In addition to this we have a centre of inversion at the origin. Each MO must belong to one or other of the irreducible representations of the group (*e.g.* ref. 31); for the

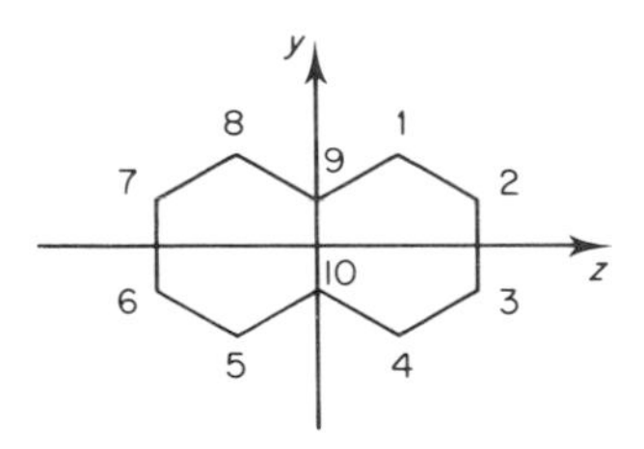

FIG. B1 Symmetry axes for the naphthalene molecule.

group, D_{2h}, where there are eight symmetry-operations, these are a_g, a_u, b_{1g}, b_{1u}, b_{2g}, b_{2u}, b_{3g}, and b_{3u}. (We use small letters for the labellings because we are going to refer initially to individual orbitals; we shall later use large letters for what happens when these orbitals are combined in conformity with the various available symmetries). This means that any molecular orbital for naphthalene—whether it be a σ-orbital or a π-one—must have the symmetry associated with one or other of these eight possibilities. Now we can sort these out a little because we can distinguish between σ and π. Chemically, this distinction is associated with the fact that the σ-orbitals are unchanged by reflection in the plane of the molecule whereas the π-orbitals are multiplied by -1 on reflection. In fact, standard methods of group theory and study of character tables show that four of the above symmetries,

$$a_g, b_{1g}, b_{2u} \text{ and } b_{3u}, \tag{B1}$$

describe the symmetries of the σ-type MO's. Those which describe the symmetries of the π-type MO's are then the remainder—that is to say:

$$a_u, b_{1u}, b_{2g} \text{ and } b_{3g}. \tag{B2}$$

If we were dealing with both σ- and π-orbitals, we should, of course, need to take account of both these groups of symmetries. At the moment, however, we are concerned only with the π-orbitals and hence will use only the symmetries listed in (B2).

We now consider how this can be built into our description of the molecular orbitals which we are going to form out of the ten $2p_\pi$-atomic-orbitals on the carbon-atoms of naphthalene; these $2p_\pi$ AO's form a basis for representing the group, D_{2h}—they are, as-it-were, the "bricks" out of which the final π-MO's are to be constructed. If we take these ten $2p_\pi$-orbitals and then ask what happens to them under the various rotations and reflections, and inversion, of the group, it is possible to see very quickly that these operations can only interchange these ten $2p_\pi$-orbitals—we cannot introduce anything new by this process. It is for this reason that we say these orbitals form a *basis* for the representation, *i.e.* the irreducible representations, of the group. By standard, group-theoretical methods, we find that, of the ten π-MO's, two have symmetry a_u, three have b_{1u}, two have b_{2g} and three are of b_{3g}-symmetry—*i.e.*

$$2a_u + 3b_{1u} + 2b_{2g} + 3b_{3g} \tag{B3}$$

The simplest way to form these π-MO's is to start with the ten π-AO's and consider forming the correct linear-combinations of them which possess the required symmetries. Let us begin by thinking, for the moment, about ϕ_1, the AO on atom 1 (Fig. B1); now any rotation or reflection, or indeed any of the other operations of the group, will either leave this unchanged or turn it into an orbital centred on atoms 4, 5 or 8 (Fig. B1)—this is obviously so since 1, 4, 5, and 8 are symmetrically "equivalent" atoms. Therefore, whatever we do amongst $\{\phi_1, \phi_4, \phi_5, \phi_8\}$, whatever we do with the various symmetry-operations in connection with this set, can only mix them together. These four orbitals may be taken together in four different allowed-combinations of distinct symmetries:

$$\left.\begin{array}{ll} \phi_1 + \phi_4 + \phi_5 + \phi_8 & \text{symmetry } b_{1u} \\ \phi_1 + \phi_4 - \phi_5 - \phi_8 & \text{symmetry } b_{2g} \\ \phi_1 - \phi_4 - \phi_5 + \phi_8 & \text{symmetry } b_{3g} \\ \phi_1 - \phi_4 + \phi_5 - \phi_8 & \text{symmetry } a_u \end{array}\right\} \tag{B4}$$

Now these are not, of course, complete molecular-orbitals—that will come later; but these *are* the allowed symmetry-adapted combinations of the equivalent atomic-orbitals, ϕ_1, ϕ_4, ϕ_5, ϕ_8. Therefore, when we are, in a sense, ready to start building the complete molecular-orbitals, we know that in, for example, the MO's of b_{2g}-symmetry, the particular set of four orbitals $\{\phi_1, \phi_4, \phi_5, \phi_8\}$ occurs *only* in the form $(\phi_1 + \phi_4 - \phi_5 - \phi_8)$ and *not* in any one of the other three combinations listed in (B4).

By an exactly similar line of reasoning, we may show that the set of orbitals, $\{\phi_2, \phi_3, \phi_6, \phi_7\}$, centred on the equivalent carbon-atoms 2, 3, 6 and 7 (Fig. B1) may combine as follows, with symmetries as indicated:

$$\left.\begin{array}{ll} \phi_2 + \phi_3 + \phi_6 + \phi_7 & \text{symmetry } b_{1u} \\ \phi_2 + \phi_3 - \phi_6 - \phi_7 & \text{symmetry } b_{2g} \\ \phi_2 - \phi_3 - \phi_6 + \phi_7 & \text{symmetry } b_{3g} \\ \phi_2 - \phi_3 + \phi_6 - \phi_7 & \text{symmetry } a_u \end{array}\right\} \tag{B5}$$

Finally, 9 and 10 form a set of equivalent atoms on their own. These can only go together symmetrically as

$$\left.\begin{array}{ll} \phi_9 + \phi_{10} & \text{symmetry } b_{1u} \\ \text{or antisymmetrically as} & \\ \phi_9 - \phi_{10} & \text{symmetry } b_{3g} \end{array}\right\} \tag{B6}$$

This means that if, for example, we were interested in molecular orbitals, of say, b_{3g}-symmetry, these would have to be made up of a certain combination of $(\phi_1 - \phi_4 - \phi_5 + \phi_8)$ plus a certain multiple of $(\phi_2 - \phi_3 - \phi_6 + \phi_7)$ plus a certain combination of $(\phi_9 - \phi_{10})$. Furthermore, there are three different ways in which these multiples themselves may be combined and this will give rise to three distinct π-MO's of b_{3g}-symmetry—as was anticipated in scheme (B3). In all, by appropriately combining (B4), (B5) and (B6), we see that there are three-component combinations of symmetries b_{1u} and b_{3g}, and two each of b_{2g} and a_u. By using the combinations (B4), (B5) and (B6) as a basis, we thus reduce the (10×10)-determinant which we would have had to handle if we had done the whole job without bothering about symmetry, to two smaller determinants of order 3×3 (for the combinations of b_{1u}- and b_{3g}-symmetry) and two others of order 2×2 (for the a_u- and b_{2g}-symmetry-orbitals).

Clearly, for one thing, this represents a great saving in labour; a determinant of order 10×10 is really quite a handful to cope with; of course, these days, we could handle this by means of a computer but if we did not have a computer available we should have to evaluate the eigenvalues and eigenvectors of a (10×10)-matrix if the methods of group theory are not invoked—a formidable task!

A second advantage in the use of group theory is that it gives information concerning both the possible symmetries of the final π-MO's and the characterisation of subsequent transitions which might occur from the ground state to the various excited-states. This is a whole discussion in itself and we will simply conclude it here by writing down the ground state of naphthalene. The Hückel energies which result from this sort of analysis are shown in Fig. B2. In the ground state, all the bonding orbitals are doubly filled and all the anti-bonding orbitals are empty. Thus, the ground state would be described by (Fig. B2):

$$(1b_{1u})^2(1b_{2g})^2(1b_{3g})^2(2b_{1u})^2(1a_u)^2 \qquad \text{symmetry } {}^1A_g \tag{B7}$$

Since this a completely symmetrical configuration with respect to the molecular framework, it is denoted ${}^1A_g$—a *large* letter being used now because it refers to a symmetry classification involving the *whole molecule*, and not just a particular MO. Electronic transitions in which one or more electrons are excited from bonding orbitals to anti-bonding ones can now be discussed and an interpretation can be given for the ultra-violet spectrum of this molecule—indeed, the major transitions of all molecules of this sort are now categorised in this way.

We may also note in passing that the Coulson–Rushbrooke "Pairing"-Theorem for the energy levels of alternant hydrocarbons (much discussed in Chapter Six and in Appendix D) is nicely illustrated by Fig. B2 (*cf* also

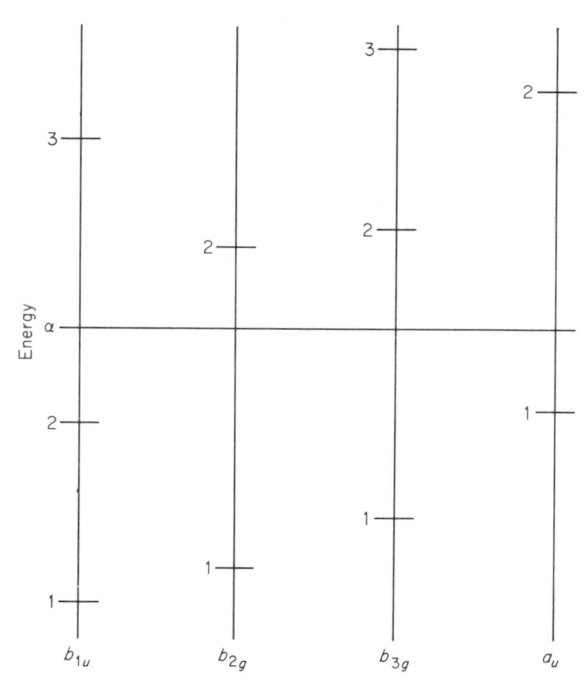

FIG. B2 π-Electron energy-levels of naphthalene, classified according to the symmetries of the corresponding molecular-orbitals. (Redrawn from C. A. Coulson's Chapter "π-Bonds" in *Physical Chemistry. An Advanced Treatise*, Vol. V. *Valency* (Editors, H. Eyring, D. Henderson, and W. Jost), Academic Press, New York, San Francisco, London, 1970, p. 415.)

Fig. 6-7). We can even discuss the symmetries of MO's which occur in complementary pairs; hence, it is seen from Fig. B2 that MO's of symmetry b_{3g} are the conjugate pairs of MO's of symmetry b_{1u} and that b_{2g}- and a_u-orbitals are similarly related.

B3 Concluding Remarks

The reduction of the full secular-determinant to a product of smaller ones, and the association of each of these smaller ones with particular symmetries, provides a powerful means of simplification, and also of interpretation. The example that we have used to illustrate it is quite typical, and represents a major simplification introduced by group theory.

Appendix C

A Matrix Proof of the Annulene Energy-Level Formula

C1 The Properties of Circulant Matrices

The simplest and most elegant proof of the energy-level formula for annulenes (equation (5-2)) exploits the properties of *circulant matrices.* A *circulant* is a matrix of the form

$$\begin{pmatrix} a_1 & a_2 & a_3 & \cdots & a_n \\ a_n & a_1 & a_2 & \cdots & a_{n-1} \\ a_{n-1} & a_n & a_1 & \cdots & a_{n-2} \\ \vdots & \vdots & \vdots & & \vdots \\ a_2 & a_3 & a_4 & \cdots & a_1 \end{pmatrix}$$

A circulant matrix is thus completely defined once the elements in its first row have been specified; furthermore, the eigenvalues and eigenvectors of a circulant may be obtained analytically in closed form, as we shall now show.

First consider the quantity

$$\lambda_k = a_1 + a_2\omega_k + a_3\omega_k^2 + \cdots + a_n\omega_k^{n-1} \tag{C1}$$

where ω_k is an nth root of unity—*i.e. one* of the n roots of the scalar equation

$$\omega^n = 1 \tag{C2}$$

which has solutions $\{\omega_k\}$, $k = 1, 2, \ldots, n$,

$$\omega_k = \cos\left(\frac{2k\pi}{n}\right) + i\sin\left(\frac{2k\pi}{n}\right) \tag{C3}$$

Evidently, from equation (C1), it is clear that the following equations are true:

$$\left.\begin{array}{r}\lambda_k = a_1 + a_2\omega_k + a_3\omega_k^2 + \cdots + a_n\omega_k^{n-1} \\ \omega_k\lambda_k = a_n + a_1\omega_k + a_2\omega_k^2 + \cdots + a_{n-1}\omega_k^{n-1} \\ \cdots\cdots\cdots\cdots\cdots\cdots\cdots\cdots\cdots\cdots\cdots\cdots \\ \omega_k^{n-1}\lambda_k = a_2 + a_3\omega_k + a_4\omega_k^2 + \cdots + a_1\omega_k^{n-1}\end{array}\right\} \tag{C4}$$

and the set of equations (C4) implies that

$$\begin{pmatrix} a_1 & a_2 & a_3 & \cdots & a_n \\ a_n & a_1 & a_2 & \cdots & a_{n-1} \\ a_{n-1} & a_n & a_1 & \cdots & a_{n-2} \\ \vdots & \vdots & \vdots & & \vdots \\ a_2 & a_3 & a_4 & \cdots & a_1 \end{pmatrix} \begin{bmatrix} 1 \\ \omega_k \\ \omega_k^2 \\ \vdots \\ \omega_k^{n-1} \end{bmatrix} = \lambda_k \begin{bmatrix} 1 \\ \omega_k \\ \omega_k^2 \\ \vdots \\ \omega_k^{n-1} \end{bmatrix} \tag{C5}$$

Hence, the vector

$$\mathbf{V}_k = \begin{bmatrix} 1 \\ \omega_k \\ \omega_k^2 \\ \vdots \\ \omega_k^{n-1} \end{bmatrix} \tag{C6}$$

is an eigenvector of the circulant matrix under consideration, with eigenvalue λ_k, given by equation (C1). Since ω_k is just one of the n roots of (C2), it follows that there will be a $\mathbf{V}_k$ (equation (C6)) and a λ_k (equation (C1)) corresponding to every distinct ω_k, $k = 1, 2, \ldots, n$. $\{\mathbf{V}_k\}$, $k = 1, 2, \ldots, n$, and $\{\lambda_k\}$, $k = 1, 2, \ldots, n$, are thus the complete set of eigenvectors and corresponding eigenvalues of the circulant matrix in question. Note that the set of eigen*vectors*, $\{\mathbf{V}_k\}$, is the same for *all* circulant matrices, but that the eigen*values* are dependent explicitly on the elements which comprise any row of the particular circulant matrix under investigation.

C2 Application to Annulenes

The Hückel Hamiltonian-matrix for an n-annulene (Fig. 5-1) is the matrix in the determinant found in equation (5-1), with the quantities $(-x)$, along the diagonal, replaced by zero (*i.e.* α). As pointed out in Appendix A, this matrix is isomorphic with the adjacency matrix, $\mathbb{A}(C_n)$, of the correspondingly-labelled molecular graph; furthermore, the eigenvalues of the matrix, $\mathbb{A}(C_n)$, are the annulene energy-levels we require (expressed in *units* of β and *relative* to α). Now this adjacency matrix, $\mathbb{A}(C_n)$, of the cyclic graph representing the carbon-atom connectivity of an n-annulene *is a circulant matrix, with first row* $(0, 1, 0 \cdots 0, 1)$. In the terminology of the matrix in equation (C5), therefore, only a_2 and a_n (both unity) are non-zero. Immediately, therefore, from equation (C1),

$$\begin{aligned} \lambda_k &= \omega_k + \omega_k^{n-1} \\ &= \omega_k + \omega_k^* \\ &= 2\cos\left(\frac{2k\pi}{n}\right) \end{aligned} \tag{C7}$$

the last step being *via* equation (C3). Thus, the eigenvalues of the matrix $\mathbb{A}(C_n)$ comprise the $\{\lambda_k\}$, $k = 1, 2, \ldots, n$, given by equation (C7). Naturally enough, since $\mathbb{A}(C_n)$ is a real-symmetric matrix, these eigenvalues turn out to be entirely real. Finally, since we have agreed that in this discussion energies shall be measured in *units* of β and with *reference* to α, *i.e.*

$$\lambda_k = \frac{\varepsilon_k - \alpha}{\beta} \tag{C8}$$

it follows from (C7) and (C8) that

$$\varepsilon_k = \alpha + 2\beta\cos\left(\frac{2k\pi}{n}\right) \tag{C9}$$

which is equation (5-2) for the kth energy-level of an n-annulene.

R. B. Mallion and B. O'Leary

Appendix D

A Theorem in Matrix Algebra Embodying Parts 1 and 2 of the Coulson–Rushbrooke Theorem

D1 A Matrix Theorem

Consider any $n \times n$ matrix, $\mathbb{M}$, which can be partitioned into blocks in the following way:

$$\mathbb{M} = \begin{array}{c} \uparrow \\ m \\ \downarrow \\ \uparrow \\ n-m \\ \downarrow \end{array} \left(\begin{array}{c|c} \overset{\longleftarrow m \longrightarrow}{\mathbb{O}} & \overset{\longleftarrow n-m \longrightarrow}{\mathbb{U}} \\ \hline \mathbb{V} & \mathbb{O} \end{array} \right) \tag{D1}$$

We can then state and prove the following theorem:

Theorem: If the vector

$$\mathbf{C}_I = \begin{pmatrix} c_{I1} \\ c_{I2} \\ \vdots \\ c_{Im} \\ c_{I,m+1} \\ c_{I,m+2} \\ \vdots \\ c_{In} \end{pmatrix}$$

is an eigenvector of $\mathbb{M}$ with eigenvalue λ_I, then the vector

$$\mathbf{C}'_I = \begin{pmatrix} c_{I1} \\ c_{I2} \\ \vdots \\ c_{Im} \\ -c_{I,m+1} \\ -c_{I,m+2} \\ \vdots \\ -c_{In} \end{pmatrix}$$

is an eigenvector of $\mathbb{M}$ with eigenvalue $-\lambda_I$.

Proof: If the vector, $\mathbf{C}_I$, is an eigenvector of $\mathbb{M}$ with eigenvalue λ_I then, from the definition of these quantities, this means that

$$\begin{array}{c} \\ m \\ \\ n-m \end{array} \left(\begin{array}{c|c} \leftarrow m \rightarrow & \leftarrow n-m \rightarrow \\ \mathbb{O} & \mathbb{U} \\ \hline \mathbb{V} & \mathbb{O} \end{array}\right) \begin{pmatrix} c_{I1} \\ c_{I2} \\ \vdots \\ c_{Im} \\ c_{I,m+1} \\ c_{I,m+2} \\ \vdots \\ c_{In} \end{pmatrix} = \lambda_I \begin{pmatrix} c_{I1} \\ c_{I2} \\ \vdots \\ c_{Im} \\ c_{I,m+1} \\ c_{I,m+2} \\ \vdots \\ c_{In} \end{pmatrix} \tag{D2}$$

Performing the multiplication on the left-hand side of (D2) gives

$$\left(\begin{array}{c} \mathbb{U} \begin{pmatrix} c_{I,m+1} \\ c_{I,m+2} \\ \vdots \\ c_{In} \end{pmatrix} \\ \hline \mathbb{V} \begin{pmatrix} c_{I1} \\ c_{I2} \\ \vdots \\ c_{Im} \end{pmatrix} \end{array}\right) = \lambda_I \left(\begin{array}{c} c_{I1} \\ c_{I2} \\ \vdots \\ c_{Im} \\ \hline c_{I,m+1} \\ c_{I,m+2} \\ \vdots \\ c_{In} \end{array}\right) \tag{D3}$$

Now consider the second vector, $\mathbf{C}'_I$, and multiply this on the left by $\mathbb{M}$

$$\begin{array}{c} \uparrow \\ m \\ \downarrow \\ \uparrow \\ n-m \\ \downarrow \end{array}\left(\begin{array}{c|c} \multicolumn{1}{c}{\longleftarrow m \longrightarrow} & \multicolumn{1}{c}{\longleftarrow n-m \longrightarrow} \\ \mathbb{O} & \mathbb{U} \\ \hline \mathbb{V} & \mathbb{O} \end{array}\right)\begin{pmatrix} c_{I1} \\ c_{I2} \\ \vdots \\ c_{Im} \\ -c_{I,m+1} \\ -c_{I,m+2} \\ \vdots \\ -c_{In} \end{pmatrix} = \begin{pmatrix} \mathbb{U} \begin{pmatrix} -c_{I,m+1} \\ -c_{I,m+2} \\ \vdots \\ -c_{In} \end{pmatrix} \\ \mathbb{V} \begin{pmatrix} c_{I1} \\ c_{I2} \\ \vdots \\ c_{Im} \end{pmatrix} \end{pmatrix} \tag{D4}$$

Comparison of the right-hand side of (D4) with equation (D3) reveals that

$$\begin{aligned} \begin{pmatrix} \mathbb{U} \begin{pmatrix} -c_{I,m+1} \\ -c_{I,m+2} \\ \vdots \\ -c_{In} \end{pmatrix} \\ \mathbb{V} \begin{pmatrix} c_{I1} \\ c_{I2} \\ \vdots \\ c_{Im} \end{pmatrix} \end{pmatrix} &= \lambda_I \begin{pmatrix} -c_{I1} \\ -c_{I2} \\ \vdots \\ -c_{Im} \\ c_{I,m+1} \\ c_{I,m+2} \\ \vdots \\ c_{In} \end{pmatrix} \\ &= (-\lambda_I) \begin{pmatrix} c_{I1} \\ c_{I2} \\ \vdots \\ c_{Im} \\ -c_{I,m+1} \\ -c_{I,m+2} \\ \vdots \\ -c_{In} \end{pmatrix} \end{aligned} \tag{D5}$$

(D4) and (D5) then give

$$\left(\begin{array}{c|c} \mathbb{O} & \mathbb{U} \\ \hline \mathbb{V} & \mathbb{O} \end{array}\right)\begin{pmatrix} c_{I1} \\ c_{I2} \\ \vdots \\ c_{Im} \\ -c_{I,m+1} \\ -c_{I,m+2} \\ \vdots \\ -c_{In} \end{pmatrix} = (-\lambda_I)\begin{pmatrix} c_{I1} \\ c_{I2} \\ \vdots \\ c_{Im} \\ -c_{I,m+1} \\ -c_{I,m+2} \\ \vdots \\ -c_{In} \end{pmatrix} \tag{D6}$$

which, by definition, means that $\mathbf{C}'_I$ is an eigenvector of $\mathbb{M}$ with eigenvalue $(-\lambda_I)$. Hence the Theorem is proved.

D2 Relationship to Parts 1 and 2 of the Coulson–Rushbrooke Theorem

We have just proved that *any* matrix which can be partitioned as the matrix $\mathbb{M}$ in equation (D1) has a spectrum of eigenvalues which is symmetric about zero and that the eigenvectors corresponding to these *complementary* eigenvalues differ only in the *sign* of the last $(n - m)$ components ("coefficients"). It is now a simple matter to show that this is the basis of parts 1 and 2 of the Coulson–Rushbrooke Theorem.

Consider the Hückel Hamiltonian-matrix, $\mathbb{H}$, for an alternant hydrocarbon, constructed on the basis of the simple Hückel-approximations (equations (6-2)–(6-5)); as an example, the matrix, $\mathbb{H}$, for butadiene is shown in equation (2-54). If the (m) starred atoms are labelled from $1, 2, \ldots, m$, and the unstarred ones from $m + 1, m + 2, \ldots, n$, then by an exactly similar argument to that used when discussing the corresponding secular-determinant for an alternant hydrocarbon in §6.3, the matrix $\mathbb{H}$ may be partitioned as in equation (D7).

$$\mathbb{H} = \begin{array}{c} \scriptstyle \uparrow \, m \, \downarrow \\[2em] \scriptstyle \uparrow \, n-m \, \downarrow \end{array} \overset{\xleftarrow{\;m\;}\xrightarrow{} \quad \xleftarrow{\;n-m\;}\xrightarrow{}}{\left(\begin{array}{c|c} \begin{matrix} \alpha & & & \mathbb{O} \\ & \alpha & & \\ & & \ddots & \\ \mathbb{O} & & & \alpha \end{matrix} & \textcircled{β} \\ \hline \textcircled{$\tilde{\beta}$} & \begin{matrix} \alpha & & & \mathbb{O} \\ & \alpha & & \\ & & \ddots & \\ \mathbb{O} & & & \alpha \end{matrix} \end{array}\right)} \tag{D7}$$

where the symbols $\textcircled{$\beta$}$ and $\textcircled{$\tilde{\beta}$}$ have the same meaning as they had in §6.3. If we now agree (as we have often done) to measure all energies *relative* to α, and in *units* of β, *i.e.*,

$$x_I = \frac{\varepsilon_I - \alpha}{\beta} \tag{D8}$$

$\mathbb{H}$ becomes (when divided by β)

$$\mathbb{H} = \begin{array}{c} \uparrow \\ m \\ \downarrow \\ \uparrow \\ n-m \\ \downarrow \end{array} \left(\begin{array}{c|c} \overset{\leftarrow m \rightarrow}{\mathbb{O}} & \overset{\leftarrow n-m \rightarrow}{\mathbb{B}} \\ \hline \tilde{\mathbb{B}} & \mathbb{O} \end{array} \right) \tag{D9}$$

where

$$\left.\begin{aligned} \mathbb{B} &= \frac{1}{\beta} \textcircled{\beta} \\ \text{and} \qquad & \\ \tilde{\mathbb{B}} &= \frac{1}{\beta} \textcircled{\tilde{\beta}} \end{aligned}\right\} \tag{D10}$$

$\mathbb{H}$ in equation (D9) is now partitioned exactly as $\mathbb{M}$ in equation (D1), with $\mathbb{U} = \mathbb{B}$ and $\mathbb{V} = \tilde{\mathbb{B}}$. Hence, $\mathbb{H}$ will have a set of eigenvalues $\{x_I\}$, $I = 1, 2, \ldots,$ n, which is *symmetrical about the chosen zero*—in this case, α. This is part 1 of the Coulson–Rushbrooke Theorem, (otherwise proved in §6.3).

Part 2 of the Theorem follows immediately from our previous observation (§6·5 and Appendix A) that the eigenvectors of $\mathbb{R}$ are the LCAO-coefficients of the various MO's whose energies are associated with the corresponding eigenvalues of $\mathbb{H}$. Now by the matrix theorem just proved, the eigenvectors giving rise to two complementary eigenvalues will differ only in the *sign* of the last $(n - m)$ coefficients—and so, therefore, will the LCAO-MO's associated with complementary energy-levels of an alternant hydrocarbon. This is part 2 of the Coulson–Rushbrooke Theorem, otherwise proved in §6.4.

D3 Conditions Required for the Coulson–Rushbrooke Theorem to Hold

When parts of 1 and 2 of the Coulson–Rushbrooke Theorem are proved as in this Appendix, the necessity of the condition stipulated by equation (6-2) (all Coulomb integrals, α_r, to be the same (α), for $r = 1, 2, \ldots, n$) is even more immediately evident than it was in the previous proof (§6.3). If *just one* of the diagonal elements of $\mathbb{H}$ were different from α (*i.e.*, when in the form of equation (D9), different from *zero* since, in this latter equation, the common Coulomb-integral, α, is chosen as the energy zero) then this would spoil the partitioning of the matrix into zero blocks in its upper-left and lower-right corners and hence the Pairing Theorem would no longer hold. It is also clear (as it was in the proof given in §§6.3 and 6.4) that we require

$$S_{rs} = \delta_{rs}$$

and

$$H_{rs} = 0, r \neq s \qquad r, s \textit{ non}\text{-neighbours} \tag{D11}$$

for the Theorem to hold; however, the condition that all non-zero H_{rs}, $r \neq s$ (between bonded pairs of atoms, r and s) be the same (β) is *not* required since, in the above arguments (§D2), it was not necessary to specify what the elements of $\mathbb{B}$ and $\tilde{\mathbb{B}}$ actually were. If, however, this restriction *is* made, then all the elements of $\mathbb{B}$ are either 0 or 1—and the matrix $\mathbb{H}$ then in fact becomes identical to the adjacency matrix, $\mathbb{A}$, of the (similarly labelled) graph representing the carbon-atom connectivity of the molecule in question. (See Appendix A and §D4 of this Appendix).

D4 Relationship to other Matrix Theorems and to Graph Theory

We now see that parts 1 and 2 of the Coulson–Rushbrooke Theorem are really a special case of the matrix theorem proved in §D1; in fact, this matrix theorem is, in its turn, only a special case of the famous Perron–Frobenius Theorem on matrices with non-negative elements (1907–1912)[R47].

We can also relate the Coulson–Rushbrooke Theorem to the theory of abstract graphs outlined in Appendix A. A graph in which the vertices can be divided into two groups such that no two vertices of the same group are adjacent (*i.e.*, are joined by an edge) is said to be a *bipartite graph*; the carbon-atom connectivity of an alternant hydrocarbon can thus be represented as a bipartite graph, while that of a non-alternant hydrocarbon would constitute a *non-bipartite* graph. It is thus possible to partition the adjacency matrix, $\mathbb{A}$, (see Appendix A) of an appropriately labelled bipartite graph in the way in which the matrix $\mathbb{M}$ is partitioned in equation (D1)—in fact, $\mathbb{A}$ would belong to an even more-limited class of matrices of this general type than does $\mathbb{H}$; for $\mathbb{A}$ would be of the form of $\mathbb{H}$ in equation (D9) but with the further restriction that the elements of $\mathbb{B}$ (and thus of $\tilde{\mathbb{B}}$) are all either 0 or 1. We should therefore certainly expect a theorem to exist in abstract graph theory to the effect that the eigenvalues of the adjacency matrix of a bipartite graph occur in complementary pairs, with eigenvectors related in the way we have described; indeed, there *is* such a theorem[R48]—but (independent) proofs of it appeared in the graph-theoretical literature only *after* the Coulson–Rushbrooke (1940) proof[R49] which was fashioned in the more "chemical" context of Hückel theory. As we have seen, once the transition has been made from a given graph to an adjacency matrix of it, the "Pairing Theorem" is implicit in the long-known Perron–Frobenius Theorem; even so, this is one of those apparently rare instances in which a theorem which is essentially a theorem in Pure Mathematics (in this case graph theory) was proposed and proved by physical scientists, working in a chemical context, *before* it was independently discovered by mathematicians *per se*[R50].

D5 Final Comments on the "Topological" Nature of the Coulson–Rushbrooke Theorem

We thus now appreciate the rather remarkable fact that parts 1 and 2 of the Coulson–Rushbrooke Theorem have an underlying mathematical significance outside the context of, not merely the HMO-model, but Quantum Theory, and indeed, Chemistry itself. Part 3, however, is, on the face of it, distinctly more "chemical" (or, at least, "physical") in nature. It talks of "carbon-atom charges"—a concept without immediate counterpart in graph theory (unlike "MO energy-level" and "LCAO-MO" which, when the simplest HMO-approximations are invoked, correspond precisely, as we have seen (Appendix A), to "eigenvalue" and "eigenvector", respectively, of the associated graph.) Implicitly involved also in the concept of charge is the *Aufbau* Principle, involving the Pauli Exclusion Principle and Hund's Rules for assigning electrons to the available orbitals; these likewise have no immediately obvious graph-theoretical counterparts. It appears at first sight, therefore, that part 3 of the Theorem has a much more physical, rather than purely mathematical, basis. On the other hand, although it is perhaps stretching the argument slightly, we could also regard atomic charges (as defined, for example, in equation (4-4))—at least in neutral hydrocarbons—as a purely "topological" quantity, dependent solely on carbon-atom connectivity. For if we were to agree to accept the *Aufbau* Principle, "God-given", as-it-were, purely as an abstract "prescription" for obtaining the v_I-terms of equation (4-4), then q_r would depend only on $\{c_{Ir}\}$, $I = 1, 2, \ldots, n$. These latter terms are themselves a function only of carbon-atom connectivity and hence we could regard q_r as being similarly dependent. Notice that in the last sentence, we were careful to say "a function ... of carbon-atom connectivity" rather than "a function of $\mathbb{A}$"; for although $\mathbb{A}$ (equation (A1)) certainly *is* a function of carbon-atom connectivity it is a more restricted one than we require for the Coulson–Rushbrooke Theorem since all its non-zero off-diagonal elements are precisely unity. This is equivalent to requiring that all non-zero off-diagonal elements of $\mathbb{H}$ shall be exactly β—a requirement which, as we have seen (§§6.6 and D3), is not necessary for the validity of any part of the Coulson–Rushbrooke Theorem.

R. B. Mallion and B. O'Leary

Notes

1. From his own card-index, which he considered more accurate for the earlier period (though less so for later stages in the development of the subject), Coulson felt that the number of papers published in the 1930's might be twice that suggested by A. Streitwieser in the preface to *Molecular-Orbital Theory for Organic Chemists*, Wiley, New York, 1961.
2. Since Coulson made that statement (in 1971), the VB-approaches (at least the resonance-theory versions of them) have enjoyed a recent, albeit limited, revival. *E.g.* W. C. Herndon (a) *J. Amer. Chem. Soc.*, **95**, 2404 (1973); (b) *Tetrahedron*, **19**, 3 (1973) (and subsequent papers by Herndon and his collaborators, and by other groups).
3. In this book molecular orbitals will always be denoted by upper-case letters and lower-case letters will refer to the numbering of the atoms in the particular conjugated system in question. For example, c_{Js} means "the coefficient of the atomic orbital centred on atom s in the Jth molecular orbital".
4. In this book all operators will be denoted by script letters.
5. As Coulson, with characteristic humour, observed to his undergraduate audience during this lecture, a "constant refrain" in Hückel theory is "It's the best that we can do"!
6. In expanding the first line of equation (2-15) use has been made of a very general principle in Quantum Theory which arises from the fact that any Hamiltonian operator is self-adjoint, or Hermitian. In the terminology of equation (2-15) (ϕ_i's real), this means that:

$$\int \phi_r \mathscr{H} \phi_s \, d\tau = \int \phi_s \mathscr{H} \phi_r \, d\tau \qquad \text{(2-15a)}$$

   Hence, the summations involving the non-diagonal terms in the second and third lines of equation (2-15) are over $r < s$, and the resulting factor of 2 is taken outside the summation sign.
7. The reader is reminded at this stage that, in equations (2-26) and (2-27), the quantities ($c_1, c_2, \ldots, c_n$) are the *variables* and the quantities, H_{ij} and H_{ii}, are in principle, *known* and *constant* (since the latter depend only on the 2p-type basis-atomic-orbitals, $\{\phi_i\}$, and the effective, one-electron Hamiltonian, $\mathscr{H}$.) y in equation (2-26)—replaced by ε in equations (2-27) and (2-28)—is also a variable. According to equation (2-28), we seek values of ε which make the determinant on the left-hand side of that equation equal to zero.

8. Notice that this situation is one which arises quite naturally in any eigenvalue-problem. Since

$$\mathscr{H}\Psi_I = \varepsilon_I \Psi_I \tag{2-32a}$$

is the equation which Ψ_I and ε_I satisfy, and since, from (2-31), this can be written

$$\mathscr{H}\left(\sum_r c_{Ir}\phi_r\right) = \varepsilon_I\left(\sum_r c_{Ir}\phi_r\right) \tag{2-32b}$$

it follows that each $\{c_{Ir}\}$, $r = 1, 2, \ldots, n$, may be multiplied by any arbitrary, scalar, constant-factor and, because $\sum_r(c_{Ir}\phi_r)$ occurs on *both* sides of (2-32b), this latter equation still holds. Hence, only the *relative magnitudes* $c_{I1}:c_{I2}:c_{I3}\cdots:c_{In}$ are determined by this procedure.
9. Less pictorially, and more rigorously, we ask the more mathematically-minded reader to realise that ϕ_r^σ is an even function over the integration domain while ϕ_s^π is an odd function over this domain. The product $\phi_r^\sigma\phi_s^\pi$ is thus an odd function and the integral of an odd function over a symmetrical interval is zero.
10. Again, we can say that, with respect to the operation of reflection in the molecular plane, ϕ_r^σ is an even function, $\mathscr{H}$ is an even operator and ϕ_s^π is an odd function; the combination $\phi_r^\sigma\mathscr{H}\phi_s^\pi$ is, therefore, an odd function, overall, with respect to this reflection-operation, and so, again, the integral of an odd function over a symmetrical interval is zero. Those readers familiar with group theory will realise that this is another way of saying that if two functions belong to different irreducible-representations, the integral of their product will be zero.
11. For this reason, Hückel theory lends itself very naturally to investigation by the mathematical techniques of graph theory. For brief details see Appendix A and for a more-complete review of the graph-theoretical aspects of Hückel theory, see I. Gutman and N. Trinajstić, "Graph Theory and Molecular Orbitals", *Fortschritte der Chemischen Forschung* (*Topics in Current Chemistry*) **42**, 49 (1973).
12. Note that re-numbering the vertices of the conjugated system is equivalent to permuting rows and simultaneously permuting columns of the Hückel Hamiltonian-matrix, $\mathbb{H}$, and of the secular determinant on the left-hand side of (2-58). By the properties of *zero* determinants, the value of the secular determinant for a given value of x does not change by this process. Furthermore, from the general properties of matrices, the eigenvalues of $\mathbb{H}$ (the $\{\varepsilon_I\}$ of equation (2-59)) also remain invariant under simultaneous permutation of rows and of columns of $\mathbb{H}$. The way in which the atoms of a given conjugated molecule are labelled is therefore totally arbitrary and immaterial in a calculation of this sort—as is physically reasonable since vertex-numbering is merely a convenient artifact of the calculation, introduced in order to obtain the MO-energies, $\{\varepsilon_I\}$.
13. This is a convenient point for the more mathematically-inclined reader to consider Appendix A, which illustrates the relevance of abstract *Graph Theory* (a branch of Pure Mathematics) to simple Hückel Molecular-Orbital theory; those readers not wishing to do this may continue with the text without loss of understanding.
14. It is unfortunate that in the butadiene example (which is a convenient one since the molecule is large enough to illustrate general applicability of the method and yet small enough to enable a calculation on it to be worked through manually) the two distinct absolute values which occur amongst the roots of (2-61) (1·618 and 0·618) differ by 1·000. This is purely coincidental and is in no way a general phenomenon. What is *not* coincidental about the roots of (2-61), however, is their occurrence

symmetrically in pairs ($\pm 1{\cdot}618\beta$ and $\pm 0{\cdot}618\beta$) about an appropriate reference-zero (α); this is a specific example of a general theorem (called the Coulson–Rushbrooke Pairing-Theorem) applicable to molecules of this type, which will be discussed and proved in Chapter Six.

15. Again we must point out here another unfortunate numerical-coincidence arising from the rather simple nature of the butadiene example. The LCAO-coefficients of Ψ_2 are simply permutations (with occasional changes of sign) of those of Ψ_1; this is not a general phenomenon and must in this case be considered fortuitous. The coefficients in equations (2-67) do however illustrate one pattern which *is* general for this type of molecule—a pattern which is characteristic of the LCAO-coefficients of *paired* bonding- and anti-bonding-orbitals (such as for example, Ψ_1 and Ψ_4 with energies $\alpha + 1{\cdot}618\beta$ and $\alpha - 1{\cdot}618\beta$, and Ψ_2 and Ψ_3 with energies $\alpha + 0{\cdot}618\beta$ and $\alpha - 0{\cdot}618\beta$); for such pairs of orbitals, the *absolute* values of corresponding LCAO-coefficients are the same but there is an *alternation* (on the sequential numbering used in Fig. 2-6) of sign between the several pairs of coefficients; thus, consider Ψ_2 and Ψ_3:$c_{31} = c_{21}$, but $c_{32} = -c_{22}$; and $c_{33} = c_{23}$ but $c_{34} = -c_{24}$. This is another consequence of the Coulson–Rushbrooke Pairing-Theorem, referred to in note 14 and further discussed and proved in Chapter Six.
16. When two of the $\{x_I\}$, $I = 1, 2, \ldots, n$, which are roots of an appropriate secular determinant of the form of (2-58) are identical it means that there are *two* sets of LCAO-combinatorial weighting-coefficients (*i.e.*, two combinations of the basis orbitals $\{\phi_r\}$, $r = 1, 2, \ldots, n$) *which give rise to molecular orbitals of the same energy*. (Another way of saying this, which is more directly related to the theory of real-symmetric matrices, the theory underlying the HMO-method, is to say that there are two different eigenvectors of the Hückel Hamiltonian-matrix, $\mathbb{H}$, which give rise to the same eigenvalue). Such MO's with the same energy are said to be *degenerate*. They are represented schematically on the energy-level diagrams such as Fig. 2-12 by the appropriate number of horizontal lines at the same level on the energy scale.
17. This is a figure based on the work of Mulliken, Ricke, and Brown, and of Coulson and Altmann, cited by Coulson in his chapter "π-Bonds" in *Physical Chemistry; An Advanced Treatise*, Vol. V. *Valency*, edited by H. Eyring, D. Henderson, and W. Jost, Chapter 7 (pp. 394–395), Academic Press, New York and London, 1970.
18. Although Coulson himself never, of course, used the term, the quantity p_{rs} defined in equation (4–6) is often referred to as the "Coulson Bond-Order" of the bond *r-s* (because it was introduced by him in his classic paper published in the 1939 *Proceedings of the Royal Society* (**A169**, 412)); this is in order to distinguish it from alternative definitions of bond order due to Pauling, Penney, Dirac and others. When there is no ambiguity, however, a π-bond-order is normally just referred to as such, without any further qualification, it being understood that the Coulson bond-order, as defined above, is meant.
19. This result is conventionally proved group-theoretically, although the proof is quite long (*e.g.*, E. Hückel, *Zeit. Physik*, **76**, 628 (1932)). See also J. N. Murrell, S. F. A. Kettle, and J. M. Tedder, *Valence Theory*, John Wiley and Sons, Ltd., London, 1965, pp. 268–269. A much simpler proof, only four lines long, which makes appeal to the methods of graph theory and the properties of circulant matrices, has been given by one of the Authors (R. B. Mallion, *Bulletin de la Société Chimique de France*, p. 2799 (1974)) and is reproduced in Appendix C.
20. Notice that the diagrams in Fig. 5-2 confirm the prediction of the above discussion that, when n is even, the energy levels of the [n]-annulenes *are* symmetrically

disposed about α in corresponding bonding- and anti-bonding pairs, in accordance with the Coulson–Rushbrooke Pairing-Theorem (see Chapter Six, §6.3). In 1952, A. A. Frost and B. Muslin (*J. Chem. Phys.*, **21**, 572) proposed the following simple mnemonic-device for obtaining the energy levels of an $[n]$-annulene given analytically in equation (5-2). Construct a circle of radius 2β and a vertical line at the side (Fig. 5-2a); the line is going to represent the energy scale, as in Fig. 5-2a. Then inscribe in this circle a regular polygon on n vertices *with one vertex situated on the lowest point of the circle*, as judged by its projection, horizontally, onto the vertical line originally constructed at the side; then similarly join, by a horizontal line, shown as "dotted" in Fig. 5-2a, all the other vertices of the polygon to points on the vertical line. The height of every such projection, above or below the α reference-line, then signifies the energy (in units of β) of an orbital of the $[n]$-annulene in question. If two vertices project onto the vertical line at the *same* level (as occurs (twice) in the benzene example shown in Fig. 5-2a), then this signifies the existence of a *pair* of *degenerate* MO's. It is obvious from this geometrical construction that in an $[n]$-annulene there can never be more than two MO's with the same energy.

21. The importance of *bond alternation* (in the higher annulenes, especially) is nowhere more graphically illustrated than in the π-electron magnetic properties of these systems. For rings with $n = 4p$,*infinite* paramagnetism (*i.e.*, a first-order change in energy with change in magnetic field) is predicted in the absence of bond alternation. When bond alternation is taken into account, however, Hückel theory predicts $4p$-π-electron monocyclic systems to be paramagnetic, and $(4p + 2)$-π-electron systems (for which the bond-alternation effect is not nearly so important, at least for the lower members of the series) to be diamagnetic (see J. A. Pople and K. G. Untch, *J. Amer. Chem. Soc.*, **88**, 4811 (1966)). These facts are borne out experimentally (and quite dramatically) in the proton-magnetic-resonance spectra of the recently synthesised annulenes (*e.g.*,
    (a) R. C. Haddon, V. R. Haddon, and L. M. Jackman, *Fortschritte der Chemischen Forschung* (*Topics in Current Chemistry*), **16**, 103 (1971).
    (b) F. Sondheimer, *Accounts Chem. Research*, **5**, 81 (1972).
    (c) A. G. Anastassiau, *ibid.*, **5**, 281 (1972).)
22. The "starring process" *per se* was introduced by Coulson and Rushbrooke in 1940 (C. A. Coulson and G. S. Rushbrooke, *Proc. Cambridge Philos. Soc.*, **36**, 193 (1940)) but the actual classification of all conjugated hydrocarbons as "alternant" or "non-alternant" was suggested by Coulson and Longuet-Higgins in 1947 (C. A. Coulson and H. C. Longuet-Higgins, *Proc. Royal Soc.* (*London*) **A191**, 16 (1947).)
23. As we have discussed previously (§2.7, note 12), the LCAO-MO energy-levels and coefficients are independent of the way in which the atoms of the conjugated system are (arbitrarily) labelled. In the language of graph theory, (see Appendix A), the eigenvalues and eigenvectors of a graph are *graph invariants*—"invariant", that is, to a permutation of the labelling of the vertices.
24. Coincidentally, Ψ_5 and Ψ_6 of naphthalene have the same energy as Ψ_2 and Ψ_3, respectively, of butadiene.
25. *Mathematical Note:* $\mathbb{H}$ is a real-symmetric matrix and hence, from the well-known properties of such matrices, provided that all its eigenvalues $\{\varepsilon_I\}$, $I = 1, 2, \ldots, n$ are distinct, all the eigenvectors, $\{\mathbf{C}_I\}$, $I = 1, 2, \ldots, n$, of $\mathbb{H}$ will automatically be mutually orthogonal. If any given eigenvalue of $\mathbb{H}$ is (say) m-fold degenerate, then it is always possible to take appropriate linear-combinations of the m linearly independent eigenvectors all giving rise to this m-fold degenerate eigenvalue in such

a way as to give m new eigenvectors with this eigenvalue which *are* mutually orthogonal; a similar procedure may be adopted with the eigenvectors belonging to any other sets of degenerate eigenvalues which the spectrum of $\mathbb{H}$ may contain. In all that follows, it will be implicitly assumed that this process has been carried out and that, consequently, when the $\mathbf{C}_I$ are normalised, the matrix $\mathbb{C}$ is *orthogonal.*

26. As Coulson observed at this point in the lecture, with his inimitable personification of atoms and molecules, "Well, you cannot very well expect a hydrogen atom to conveniently move out of the way to enable the NO_2^+ to come up—things don't happen quite as easily as that!".
27. We feel that Coulson's words in the lecture concerning these various semi-empirical methods are worth quoting in full: "We are rapidly getting to the stage where a dictionary is needed in order that we should know what the particular sets of approximations which are being made really are. Well, you may think I am making fun of this; I am not really. What I am saying is that it is little bit regrettable that nearly everybody [who] works in this field has his own parametrisation. It makes it a little difficult to compare the results from two authors. But within one scheme one does get, very definitely, refinements which are justified and give us more-reliable values than those which we previously got through the simple Hückel-theory for π-electrons alone".
28. As Coulson typically observed to the undergraduate audience, "But, of course, big computers don't get as much worried about that sort of thing as human beings!"
29. It is with no disrespect to the late Professor Coulson's very good friend, Professor R. S. Mulliken, that we quote Coulson's affectionate remark in the lecture: "This used to be known as the 'Baker–Nathan effect' but later, thanks to Mulliken's ability to invert names, as hyperconjugation".
30. Coulson rather humorously ended his final lecture of the series with the words: "Someone said that if you ask me a question before dinner, I am an optimist—after dinner, I am a pessimist. Well, it is not quite like that, but one must use one's common-sense [about] which kind of approximation [to] use".

References

1. N. V. Sidgwick, *The Electronic Theory of Valency*, Clarendon Press, Oxford, 1927.
2. For an account of non-bonding orbitals see H. C. Longuet-Higgins, *J. Chem. Phys.*, **18**, 265 (1950).
3. Tabulated, for example, by Streitwieser, in Chapter 7 of his book referred to in note 1.
4. C. W. Haigh, R. B. Mallion, and E. A. G. Armour, *Molec. Phys.*, **18**, 751 (1970); R. B. Mallion, *ibid*, **25**, 1415 (1973).
5. Electronegativities have been extensively tabulated, for example, by L. Pauling in *The Nature of the Chemical Bond*, Cornell University Press, Ithaca, New York, 1937.
6. Chapter 5, of A. Streitwieser, *Molecular-Orbital Theory for Organic Chemists*, Wiley, New York, 1961.
7. For an extensive bibliography, see reference 6.
   E.g. one of several papers on the subject from Pauling's group at that time is
8. L. Pauling, L. O. Brockway, and J. Y. Beach, *J. Amer. Chem. Soc.*, **57**, 2705 (1935).
9. J. C. D. Brand and D. G. Williamson, *Discussions Faraday Soc.*, **35**, 184 (1963).
10. The surviving authors have not been able to trace this reference.
11. E. C. Kooyman and E. Forenhorst, *Trans. Faraday Soc.*, **49**, 58 (1953).
12. See E. Hückel, *Zeit. Physik*, **76**, 628 (1932).
13. The e^2 state factorises to ${}^1A_1 + {}^1B_1 + {}^3A_2 + {}^1B_2$ so that there is in fact no orbitally degenerate state and the distortion is thus really of the *Renner* type. Strictly, therefore, a true Jahn–Teller distortion will occur in cyclobutadiene cation and anion but *not* in the neutral species (W. N. Lipscomb, personal communication, 1978).
14. See for example
    (a) J. A. Pople and K. G. Untch, *J. Amer. Chem. Soc.*, **88**, 4811 (1966).
    (b) R. C. Haddon, V. R. Haddon, and L. M. Jackman, *Fortschritte der Chemischen Forschung* (*Topics in Current Chemistry*), **16**, 103 (1971).
    (c) F. Sondheimer, *Accounts of Chem. Research*, **5**, 81 (1972).
    (d) A. G. Anastassiau, *ibid.*, **5**, 281 (1972).
15. B. M. Trost, G. M. Bright, C. Frihart, and D. Brittelli, *J. Amer. Chem. Soc.*, **93**, 737 (1971).
16. C. A. Coulson and R. B. Mallion, *J. Amer. Chem. Soc.*, **98**, 592 (1976).
17. See, for example, A. D. McLachlan, *Molec. Phys.*, **2**, 271 (1959).
18. G. W. Wheland and D. E. Mann, *J. Chem. Phys.*, **17**, 264 (1949).
19. W. E. Moffitt, *On the Electronic Structure of Molecules*, D. Phil. Thesis, University of Oxford, 1948.
20. A. Streitwieser, *J. Amer. Chem. Soc.*, **82**, 4123 (1960).

21. H. Kuhn, *Tetrahedron*, **19** (Supplement 2), 437 (1963).
22. J. Gayoso, "*Contribution aux Méthodes Empiriques et Semi-Empiriques de la Chimie Quantique*", *Thèse, Docteur ès Sciences, Université de Paris*, 1973.
23. C. A. Coulson and F. Wille, *Tetrahedron*, **22**, 3549 (1966).
24. J. E. Lennard-Jones, *Proc. Royal Soc.* (*London*), **A158**, 280 (1937).
25. H. C. Longuet-Higgins and L. Salem, *Proc. Royal Soc.* (*London*), **A251**, 172 (1959).
26. C. A. Coulson and A. Gołebiewski; (a) *Proc. Phys. Soc.*, **78**, 1310 (1961); (b) *Molec. Phys.*, **5**, 71 (1962).
27. G. Doggett, *Molec. Phys.*, **10**, 225 (1966).
28. G. W. Wheland and L. Pauling, *J. Amer. Chem. Soc.*, **57**, 2086 (1935).
29. *E.g.* C. K. Ingold, *J. Chem. Soc.*, 1933, p. 1120 and many references contained therein.
30. For details, see C. A. Coulson and H. C. Longuet-Higgins, *Proc. Royal Soc.* (*London*), **A191**, 39 (1947).
31. See, for example, S. L. Altmann, in *Quantum Theory* (Editor: D. R. Bates), Academic Press, New York and London, Vol. II, 1962, Chapter 2.
32. R. Hoffmann and W. N. Lipscomb, *J. Chem. Phys.*, **36**, 2179, 3489 (1962); ibid., **37**, 2872 (1962).
33. R. F. Weaver, *Some Problems in the Electronic Structure of Molecules*, D. Phil. Thesis, University of Oxford, 1971.
34. R. Pariser and R. G. Parr, *J. Chem. Phys.*, **21**, 466 and 767 (1953).
35. J. A. Pople, *Trans. Faraday Soc.*, **49**, 1375 (1953).
36. C. C. J. Roothaan, *Rev. Mod. Phys.*, **23**, 69 (1951).
37. D. R. Hartree, (a) *Proc. Cambridge Philos. Soc.*, **24**, 89 (1928); (b) *The Calculation of Atomic Structures*, John Wiley, New York, 1957.
38. C. A. Coulson, *Valence*, Second Edition, Oxford University Press, London, 1961, pp. 31–35.
39. For details and bibliography of methods involving neglect of differential overlap, see also a comprehensive series of papers by J. M. Sichel and M. A. Whitehead *Theoret. Chim. Acta*, (a) **7**, 32 (1967); (b) **11**, 220, 239, 254 and 263 (1968). For a review of semi-empirical methods overall see G. Klopman and B. O'Leary, "All-Valence-Electrons SCF-Calculations of Large Organic Molecules" *Fortschritte der Chemischen Forschung* (*Topics in Current Chemistry*), **15**, 445 (1970).
40. C. A. Coulson, *Valence* (Second Edition), Oxford University Press, London, 1961, pp. 39–41.
41. S. F. Boys, *Proc. Royal Soc.* (*London*), **A200**, 542 (1950).
42. E. Clementi, *Chem. Rev.*, **68**, 341 (1968).
43. See, for example, C. A. Coulson, "σ-Bonds" in *Physical Chemistry: An Advanced Treatise*, Vol. V. *Valency* (Editors: H. Eyring, D. Henderson and W. Jost), Academic Press, New York, San Francisco, London, 1970, pp. 341–345.
44. See
    (a) M. J. S. Dewar, *Hyperconjugation*, Ronald Press, New York, 1962.
    (b) C. A. Coulson, *Valence*, Second Edition, Oxford University Press, London, 1961, Chapter 13.
45. M. D. Newton and W. N. Lipscomb, *J. Amer. Chem. Soc.*, **89**, 4261 (1967).
46. Based very closely on pp. 413–416 of C. A. Coulson's Chapter "π Bonds", which is Chapter 7 of *Physical Chemistry: An Advanced Treatise*, Vol. V, *Valency* (edited by H. Eyring, D. Henderson, and W. Jost), Academic Press, New York and London, 1970.
47. See, for example, F. R. Gantmacher, *The Theory of Matrices*, Vol. II, Chelsea Publishing Co., New York, 1960, pp. 53–56.

48. For details and references, see: (a) R. B. Mallion, A. J. Schwenk, and N. Trinajstić, "On the Characteristic Polynomial of a Rooted Graph", in *Recent Advances in Graph Theory: Proceedings of the Second Czechoslovak Symposium on Graph Theory*, (Editor: M. Fiedler), Academia, Prague, 1975, pp. 345–350; (b) M. J. Rigby and R. B. Mallion, *J. Combinatorial Theory*, in press.
49. C. A. Coulson and G. S. Rushbrooke, *Proc. Cambridge Philos. Soc.*, **36**, 193 (1940).
50. See R. B. Mallion, *Chemistry in Britain*, **9**, 242 (1973). The matrix theorem presented in the present Appendix was proved independently on at least three occasions:
    (a) K. Ruedenberg, *J. Chem. Phys.*, **29**, 1232 (1958).
    (b) D. Cvetković, *Mathematica Biblioteka*, *Belgrade*, **41**, 193 (1969).
    (c) D. H. Rouvray, *Comptes Rend. Acad. Sci.* (*Paris*), *Série* C, **274**, 1561 (1972).

Bibliography

1 Books

Without any attempt at completeness, and with apologies in advance to the authors of any books worthy of inclusion which we have omitted, we list below a somewhat subjective and arbitrary selection of the best-known previously-published books, a significant proportion of which concerns Hückel Theory.

C. A. Coulson, *Valence*, Oxford University Press, London, 1952 [Chapter IX].

A. Pullman and B. Pullman, *Les Théories Electroniques de la Chimie Organique*, Masson et Cie, Paris, 1952.

R. Daudel, R. Lefebvre and C. Moser, *Quantum Chemistry*, Interscience, New York, 1959.

J. W. Linnett, *Wave Mechanics and Valency*, Methuen and Co. Ltd., London, 1960.

A. Streitwieser, *Molecular-Orbital Theory for Organic Chemists*, Wiley, New York, 1961.

J. D. Roberts, *Notes on Molecular-Orbital Calculations*, W. A. Benjamin Inc., New York, 1961.

F. A. Cotton, *Chemical Applications of Group Theory*, Wiley-Interscience, New York, 1963.

R. G. Parr, *The Quantum Theory of Molecular Structure*, W. A. Benjamin Inc., New York, 1963.

J. N. Murrell, S. F. A. Kettle, and J. M. Tedder, *Valence Theory*, Wiley, New York, 1965.

T. E. Peacock, *Electronic Properties of Aromatic and Heterocyclic Molecules*, Academic Press, London and New York, 1965.

C. A. Coulson and A. Streitwieser, *Dictionary of π-Electron Calculations*, W. H. Freeman and Co., San Francisco, 1965.

A. Streitwieser and J. I. Brauman, *Supplemental Tables of Molecular-Orbital Calculations*, Pergamon Press, New York, 1965.

M. W. Hanna, *Quantum Mechanics in Chemistry*, W. A. Benjamin Inc., New York, 1965.

E. Heilbronner and H. Straub, *Hückel Molecular Orbitals*, Springer-Verlag, New York, 1966.

L. Salem, *Molecular-Orbital Theory of Conjugated Systems*, W. A. Benjamin Inc., New York, 1966.

A. Liberles, *Introduction to Molecular-Orbital Theory*, Holt, Rinehart and Winston, New York, 1966.

F. L. Pilar, *Elementary Quantum-Chemistry*, McGraw-Hill Inc., New York, 1968.
R. L. Flurry, *Molecular-Orbital Theories of Bonding in Organic Molecules*, Marcel Dekker Inc., New York, 1968.
E. Heilbronner and H. Bock, *Das HMO-Modell und seine Anwendung*, Verlag Chemie, Weinheim, 1968.
G. Klopman and B. O'Leary, *Introduction to All-Valence-Electrons S.C.F. Calculations of Large Organic Molecules: Theory and Applications*, in *Topics in Current Chemistry (Fortschritte der Chemischen Forschung)*, Vol. 15, p. 445, Springer-Verlag, Berlin, 1970.
D. S. Urch, *Orbitals and Symmetry*, Penguin Ltd., Harmondsworth, 1970.
R. B. Woodward and R. Hoffmann, *The Conversation of Orbital Symmetry*, Academic Press, New York, 1970.
R. Zahradnik and J. Pančir, *HMO Energy-Characteristics*, IFI Plenum, New York and London, 1970.
C. A. Coulson, "π Bonds", Chapter 5 of *Physical Chemistry; An Advanced Treatise* (edited by H. Eyring, D. Henderson, and W. Jost), Volume V, Academic Press, London and New York, 1970.
J. N. Murrell, *The Theory of the Electronic Spectra of Organic Molecules*, Chapman and Hall, London, 1971.
P. J. Garratt, *Aromaticity*, McGraw-Hill Co. (U.K.) Ltd., Maidenhead, 1971.
J. E. Banfield, *An Introduction to Mathematical Organic Chemistry*, Gereng, 1972.
E. Heilbronner and H. Bock, *The HMO-Model and its Applications; Basis and Manipulation*, Chichester, Wiley, New York, 1976.
K. Yates, *Hückel Molecular-Orbital Theory*, Academic Press, New York and London, 1978.

2 Original Papers

The major source-papers in the early development of Hückel theory (up to 1950) are as follows:

E. Hückel, (a) *Zeit. Physik*, **70**, 204,(1931); (b) *ibid.*, **72**, 310 (1931); (c) *ibid.*, **76**, 628 (1932); (d) *ibid.*, **83**, 632 (1933).
F. London, *J. Physique Radium*, (7e *Série*), **8**, 397 (1937).
J. E. Lennard-Jones, *Proc. Royal Soc.* (*London*), **A158**, 280 (1937).
C. A. Coulson, (a) *Proc. Royal Soc.* (*London*), **A164**, 383 (1938); (b) *ibid.*, **A169**, 413 (1939); (c) *Proc. Cambridge Philos. Soc.*, **36**, 201 (1940).
C. A. Coulson and G. S. Rushbrooke, *Proc. Cambridge Philos. Soc.*, **36**, 193 (1940).
C. A. Coulson and H. C. Longuet-Higgins, (a) *Proc. Royal Soc.* (*London*), **A191**, 39 (1947); (b) *ibid.*, **A192**, 16 (1947); (c) *ibid.*, **A193**, 447 (1948); (d) *ibid.*, **A193**, 456 (1948); (e) *ibid.*, **A195**, 188 (1948).
B. H. Chirgwin and C. A. Coulson, *Proc. Royal Soc.* (*London*), **A201**, 196 (1950).
C. A. Coulson, *Proc. Cambridge Philos. Soc.* **46**, 202 (1950).

Index